ROBOT SYSTEM RELIABILITY AND SAFETY

ROBOT SYSTEM
RELIABILITY AND
SAFETY

A MODERN
APPROACH

B. S. DHILLON

CRC Press
Taylor & Francis Group
Boca Raton London New York

CRC Press is an imprint of the
Taylor & Francis Group, an **informa** business

CRC Press
Taylor & Francis Group
6000 Broken Sound Parkway NW, Suite 300
Boca Raton, FL 33487-2742

First issued in paperback 2020

© 2015 by Taylor & Francis Group, LLC
CRC Press is an imprint of Taylor & Francis Group, an Informa business

No claim to original U.S. Government works

ISBN-13: 978-1-4987-0644-5 (hbk)
ISBN-13: 978-0-367-78350-1 (pbk)

Visit the Taylor & Francis Web site at
http://www.taylorandfrancis.com

and the CRC Press Web site at
http://www.crcpress.com

This book is affectionately dedicated to my grandson, Dillon.

Contents

Preface..xv
Author.. xix

1 Introduction 1
    1.1   Background..1
    1.2   Robot System Reliability/Safety-Related Facts, Figures,
          and Examples..2
    1.3   Terms and Definitions...3
    1.4   Useful Sources for Obtaining Information on Reliability
          and Safety of Robot Systems 5
          1.4.1   Journals 5
          1.4.2   Conference Proceedings 6
          1.4.3   Books..6
          1.4.4   Technical Reports..7
          1.4.5   Data Sources 7
          1.4.6   Standards 8
          1.4.7   Organizations 9
    1.5   Scope of the Book 9
    1.6   Problems.. 10
    References 11

2 Basic Mathematical Concepts 13
    2.1   Introduction 13
    2.2   Arithmetic Mean and Mean Deviation......................... 13
          2.2.1   Arithmetic Mean.. 14
          2.2.2   Mean Deviation... 14
    2.3   Boolean Algebra Laws.. 15
    2.4   Probability Definition and Properties............................ 16
    2.5   Probability Distribution-Related Definitions................ 18
          2.5.1   Cumulative Distribution Function 18
          2.5.2   Probability Density Function............................. 19
          2.5.3   Expected Value... 19
    2.6   Probability Distributions 20
          2.6.1   Exponential Distribution 20
          2.6.2   Rayleigh Distribution 21
          2.6.3   Weibull Distribution.. 22
          2.6.4   Bathtub Hazard Rate Curve Distribution 22
    2.7   Laplace Transform Definition, Common Laplace
          Transforms, and Final-Value Theorem Laplace Transform 23
          2.7.1   Laplace Transform Definition 23

       2.7.2   Laplace Transforms of Common Functions......................24

       2.7.3   Final Value Theorem Laplace Transform 25

  2.8   Solving First-Order Differential Equations Using

        Laplace Transforms..25

  2.9   Problems..27

  References 28

3 Reliability and Safety Basics 31

  3.1   Introduction 31

  3.2   Bathtub Hazard Rate Curve 31

  3.3   General Reliability-Related Formulas..................................33

       3.3.1   Probability (or Failure) Density Function...........33

       3.3.2   Hazard Rate (or Time-Dependent Failure Rate)

              Function...34

       3.3.3   General Reliability Function 34

       3.3.4   Mean Time to Failure 35

  3.4   Reliability Configurations 37

       3.4.1   Series Network..37

       3.4.2   Parallel Network 39

       3.4.3 k-out-of-n Network...41

       3.4.4   Standby System 43

       3.4.5   Bridge Network...45

  3.5   Need for Safety and the Role of Engineers with Respect

        to Safety...47

  3.6   Classifications of Product Hazards and Common

        Mechanical Injuries 48

  3.7   Organization Tasks for Product Safety and Safety

        Management Principles...49

  3.8   Accident-Causation Theories 51

       3.8.1   The "Human Factors" Accident-Causation Theory..........51

       3.8.2   The "Domino" Accident-Causation Theory 52

  3.9   Problems..54

  References 54

4 Methods for Performing Reliability and Safety Analysis

**   of Robot Systems** 57

  4.1   Introduction 57

  4.2   Failure Modes and Effect Analysis.......................................57

  4.3   The Markov Method..59

  4.4   Fault Tree Analysis 63

       4.4.1   Probability Evaluation of Fault Trees..................65

       4.4.2   FTA Advantages and Disadvantages...................67

  4.5   Technique of Operations Review...68

  4.6   Hazard and Operability Analysis 69

  4.7   Interface Safety Analysis 69

4.8   Probability Tree Method 71
4.9   Problems 73
References 74

5   **Robot Reliability** 77
5.1   Introduction 77
5.2   Classifications of Robot Failures and Their Causes
       and Corrective Measures 77
5.3   Robot Effectiveness Dictating Factors and Robot Reliability
       Survey Results 79
5.4   Robot-Related Reliability Measures 80
       5.4.1   Mean Time to Robot-Related Problems 80
       5.4.2   Mean Time to Robot-Related Failure 81
       5.4.3   Robot Reliability 82
       5.4.4   Robot's Hazard Rate 84
5.5   Robot Reliability Analysis Methods and Models
       for Performing Robot Reliability Studies 84
5.6   Reliability Analysis of Hydraulic and Electric Robots 86
       5.6.1   Reliability Analysis of the Hydraulic Robot 86
       5.6.2   Reliability Analysis of the Electric Robot 90
5.7   Problems 93
References 94

6   **Robot Safety** 97
6.1   Introduction 97
6.2   Robot Safety: Problems and Hazards 97
6.3   Roles of Robot Manufacturers and Users in Robot Safety 99
6.4   Safety Considerations in Robot Design, Installation,
       Programming, and Operation and Maintenance Phases 101
       6.4.1   Robot Design Phase 101
       6.4.2   Robot Installation Phase 102
       6.4.3   Robot Programming Phase 103
       6.4.4   Robot Operation and Maintenance Phase 103
6.5   Robot-Related Safety Problems Causing Weak Points
       in Planning, Design, and Operation 104
6.6   Robot Safeguard Approaches 105
       6.6.1   Warning Signs 105
       6.6.2   Physical Barriers 105
       6.6.3   Intelligent Systems 106
       6.6.4   Flashing Lights 106
6.7   Common Robot Safety Features and Their Functions 106
6.8   Safety Considerations for Robotized Welding Operations 107
       6.8.1   Robotized Laser Welding Hazards and Safety
                 Measures 109

        6.8.2   Hazards and Safety Measures for Robotized
                  Gas-Shielded Arc Welding 110
        6.8.3   Hazards and Safety Measures for Robotized
                  Resistance Welding.. 111
  6.9   Problems.. 111
  References 112

7 **Robot Accidents and Analysis**... 115
  7.1   Introduction 115
  7.2   Some Examples of Robot-Related Accidents................................. 115
  7.3   Robot Accidents: Causes and Sources.. 117
        7.3.1   Study I.. 117
        7.3.2   Study II 118
  7.4   Effects of Robot-Related Accidents.. 118
  7.5   Robot-Related Accidents at Manufacturer and User Facilities......119
  7.6   Useful Recommendations to Prevent Human Injury
        by Robots 120
        7.6.1   Robot System Design... 121
        7.6.2   Supervision of Workers.. 121
        7.6.3   Training of Workers.. 121
  7.7   Methods for Performing Robot Accident Analysis..................... 122
        7.7.1   Estimating Probability of an Accident Occurrence
                  Related to the Operation of a Robot.................................. 122
        7.7.2   Root Cause Analysis... 123
                7.7.2.1   RCA Software 124
                7.7.2.2   RCA Advantages and Disadvantages 124
        7.7.3   Fault Tree Analysis 125
        7.7.4   Markov Method 127
  7.8   Problems.. 130
  References 130

8 **Robot Maintenance and Areas of Robotics Applications
  in Maintenance and Repair** 133
  8.1   Introduction 133
  8.2   Robot Maintenance-Related Needs and Maintenance Types..... 133
  8.3   Commonly Used Tools to Maintain a Robot and Measuring
        Instruments and Tooling for Periodic Robot Inspections........... 134
  8.4   Robot Diagnosis and Monitoring Approaches............................ 135
  8.5   Useful Guidelines to Safeguard Robot Maintenance
        Personnel and Safeguarding Methods for Use during
        the Robot Maintenance Process... 136
  8.6   Models for Performing Robot Maintenance Analysis 137
        8.6.1   Model I.. 137
        8.6.2   Model II 139

	8.6.3	Model III	141
	8.6.4	Model IV	144
8.7	Areas of Robotics Applications in Maintenance and Repair		145
	8.7.1	Nuclear Industry	146
	8.7.2	Highways	147
	8.7.3	Railways	148
	8.7.4	Underwater Facilities	148
	8.7.5	Power Line Maintenance	149
8.8	Problems		149
References			150

9 Human Factors and Safety Considerations in Robotics Workplaces 153

9.1	Introduction		153
9.2	Human Factors-Related Issues during the Robotic Systems' Factory Integration Process		153
	9.2.1	Safety	154
	9.2.2	Maintainability	154
	9.2.3	Worker–Machine Interface	155
	9.2.4	Selection and Training	155
	9.2.5	Work Environment	155
	9.2.6	Management	156
	9.2.7	Job Design	156
	9.2.8	Communication among Workers and between Workers and Management	156
9.3	Common Robot and Robot-Related Human Tasks		156
9.4	Rules of Robotics in Regard to Humans and Advantages and Disadvantages of Robotization with Respect to Human Factors		157
9.5	Humans at Risk from Robots and Risk-Reducing Measures to Prevent Robot-Related Human Accidents		159
9.6	Useful Guidelines to Safeguard Robot Teachers and Operators		160
9.7	Approaches for Limiting Robot Movements		161
9.8	Methods for Analysis of Safety and Human Error in Robotics Workplaces		162
	9.8.1	Deviation Analysis	162
	9.8.2	Energy Barrier Analysis	162
	9.8.3	Task Analysis	163
	9.8.4	Fault Tree Analysis	163
	9.8.5	Markov Method	165
9.9	Problems		167
References			168

10 **Robot Testing, Costing, and Failure Data** 171
    10.1   Introduction 171
    10.2   Robot Performance Testing 171
    10.3   Robot Performance Testing Methods 173
           10.3.1   Test-Robot Program Method 173
           10.3.2   Ford Method 173
           10.3.3   IPA–Stuttgard Method 174
           10.3.4   National Bureau of Standards Method 174
    10.4   Robot Reliability Testing 174
           10.4.1   Management-Related Tasks in a Robot Reliability
                    Test Plan 175
           10.4.2   Useful Guidelines for Robot Reliability
                    Demonstration Tests 176
    10.5   Robot Testing and Start-Up Safety-Related Factors 176
    10.6   Robot Project Cost 177
    10.7   Models for Estimating Robot-Related Costs 178
           10.7.1   Model I 178
           10.7.2   Model II 178
           10.7.3   Model III 179
           10.7.4   Model IV 179
           10.7.5   Model V 180
           10.7.6   Model VI 180
    10.8   Robot Life Cycle Cost Estimation Models 180
           10.8.1   Life Cycle Cost Model I 181
           10.8.2   Life Cycle Cost Model II 181
    10.9   Useful Methods for Making Financial Decisions
           about Robotization 182
           10.9.1   Method I: Life Cycle Costing 182
           10.9.2   Method II: Payback Period 182
           10.9.3   Method III: Minimum Cost Rule 183
           10.9.4   Method IV: Return on Investment 184
    10.10 Failure Data Uses with Regard to Robots and Failure
           Reporting and Documentation System for Robots 184
    10.11 Main Data Sources for Reliability 185
    10.12 Repair and Inspection Records-Related Requirements
           for Robots 187
    10.13 Problems 188
    References 188

11 **Mathematical Models for Analysis of Robot-Related**
    Reliability and Safety 191
    11.1   Introduction 191
    11.2   Model I 191
    11.3   Model II 195

11.4   Model III 198
11.5   Model IV 201
11.6   Model V 205
11.7   Model VI 207
11.8   Problems 209
References 210

**Appendix A: Bibliography—Literature on the Reliability
and Safety of Robot Systems** 211

Index 233

Preface

Today, a vast number of robots are being used to perform various types of tasks around the world. In the industrial sector, robots are often used to perform tasks such as spot-welding, materials-handling, and arc welding. A robot has to be safe and reliable. An unreliable robot may become the cause of unsafe conditions, inconvenience, high maintenance cost, and so on. As robots make use of electrical, electronic, mechanical, hydraulic, and pneumatic parts, this makes the issue of their reliability very challenging because of the many different sources for potential failures.

Needless to say, robot-system reliability and safety have become an important issue. For example, with regard to robot-system safety, a study reported that 12%–17% of the accidents in industries using advanced manufacturing technology were concerned with automated production equipment.

Although, over the years, a large number of articles have appeared in journals and conference proceedings on robot reliability, safety and associated areas, to the best of author's knowledge, since 1991, there has been no book that covers all these topics within its framework. This causes a great deal of difficulty for up-to-date information-seekers because they have to consult many different and diverse sources.

Thus, the main objective of this book is to combine all these topics into a single volume and to eliminate the need to consult many different and diverse sources in obtaining up-to-date desired information on the topics. The sources of most of the material presented are given in the reference section at the end of each chapter. This will be useful to readers if they desire to delve more deeply into a specific area or topic. The book contains a chapter on mathematical concepts and another chapter on the basics of reliability and safety considered useful to understand the content of subsequent chapters. Furthermore, another chapter is devoted to methods considered useful to analyze the reliability and safety of robot systems.

The topics covered in the book are treated in such a manner that the reader will require no previous knowledge to understand the contents. At appropriate places, the book contains examples along with their solutions, and there are numerous problems at the end of each chapter to test the reader's comprehension in the area. An extensive list of publications dating 1974–2013, directly or indirectly on robot-system reliability and safety, is provided at the end of this book to give readers a view of the intensity of developments in the area.

The book comprises 11 chapters. Chapter 1 presents the need for and historical developments in robot-system reliability and safety; robot-system reliability/safety-related facts, figures, and examples; important terms and definitions; and useful sources for obtaining information on robot-system reliability and safety. Chapter 2 reviews mathematical concepts considered useful to understand subsequent chapters. Some of the topics covered in the chapter are Boolean algebra laws, probability properties, probability distributions, and useful definitions.

Chapter 3 presents various introductory aspects of reliability and safety. Chapter 4 presents a number of methods considered useful to analyze robot-system reliability and safety—failure modes and effect analysis, the Markov method, fault tree analysis, technique of operations review, hazard and operability analysis, interface safety analysis, and the probability-tree method. Chapter 5 is devoted to robot-reliability. Some of the topics covered in the chapter are classifications of robot failures and their causes, robot-reliability measures, robot-reliability analysis methods, and hydraulic and electric robots' reliability analysis.

Chapters 6 and 7 are devoted to robot-related safety and accidents and analysis of these. Chapter 6 covers topics such as robot-related safety problems and hazards, roles of robot manufacturers and users in robot-related safety, safety considerations in robot design, installation, programming, and operation and maintenance phases; robot safeguard approaches, common robot safety features and their functions, and robot welding operations-related safety considerations. Some of the topics covered in Chapter 7 are examples of robot-related accidents, accident causes and sources, useful recommendations to prevent human injury by robots, and methods for performing robot-related accident analysis.

Chapter 8 presents various important aspects of robot-maintenance and areas of robotics applications in maintenance and repair, including robot-maintenance-related needs and maintenance types, diagnosis and monitoring approaches, useful guidelines to safeguard robot-maintenance personnel, and models for performing robot-maintenance analysis. Chapter 9 is devoted to human factors and safety considerations in robotics workplaces and covers topics such as human factors-related issues during the robotic systems' factory-integration process, common robot and human tasks, rules of robotics with regard to humans, humans at risk from robots, useful guidelines to safeguard robot operators and teachers, methods to limit robot movements, and methods for performing safety and human error analysis in robotics workplaces.

Chapter 10 is devoted to testing robots, costing, and failure data. Some of the topics covered in the chapter are testing for performance of robots, testing methods, robot-reliability testing, project cost, models for estimating robot-related costs, failure-data, failure-reporting and documentation system for robots, main data sources for reliability, and data for mean time between robot failures. Finally, Chapter 11 presents six mathematical models for analyzing robot-system reliability and safety.

The book will be useful to many individuals, including system engineers, design engineers, reliability engineers, safety engineers, engineering professionals working in the area of robotics, researchers and instructors involved with robotic systems, graduate and senior undergraduate students in system engineering, reliability engineering, and safety engineering, and engineers-at-large.

The author is deeply indebted to many individuals, including family members, colleagues, friends, and students for their invisible input. The unseen contributions of my children also are appreciated. Last, but not the least, I thank my wife, Rosy, my other half and friend, for typing this entire book and for her timely help in proofreading.

B. S. Dhillon
Ottawa, Ontario, Canada

Author

Dr. B. S. Dhillon is a professor of engineering management in the Department of Mechanical Engineering at the University of Ottawa, Ontario, Canada. He has served as chairman/director of the Mechanical Engineering Department/Engineering Management Program for over 10 years at the same institution. He is the founder of the probability distribution model named the *Dhillon Distribution/Law/Model* by statistical researchers around the world. He has published over 376 articles (i.e., 226 [70 single-authored + 156 coauthored] journals and 150 conference proceedings) on reliability engineering, maintainability, safety, engineering management, and so on. He is or has been on the editorial boards of 12 international scientific journals. In addition, Dr. Dhillon has written 42 books on various aspects of healthcare, engineering management, design, reliability, safety, and quality, published by Wiley (1981), Van Nostrand (1982), Butterworth (1983), Marcel Dekker (1984), Pergamon (1986), and so on. His books are being used in over 100 countries and many of them are translated into languages such as German, Russian, Chinese, and Persian (Iranian).

He has served as general chairman of two international conferences on reliability and quality control held in Los Angeles and Paris in 1987. Dr. Dhillon has also served as a consultant in various organizations and bodies and has many years of experience in the industrial sector. At the University of Ottawa, he has been teaching reliability, quality, engineering management, design, and related areas for over 35 years and he has also lectured in over 50 countries, having delivered keynote addresses at various international scientific conferences held in North America, Europe, Asia, and Africa. In March 2004, Dr. Dhillon was a distinguished speaker at the Conference/Workshop on Surgical Errors (sponsored by the White House Health and Safety Committee and Pentagon), held on Capitol Hill (One Constitution Avenue, Washington, DC).

Dr. Dhillon attended the University of Wales, Cardiff, Wales, U.K., where he received a BS in electrical and electronic engineering and an MS in mechanical engineering. He received a PhD in industrial engineering from the University of Windsor, Ontario, Canada.

1

Introduction

1.1 Background

Today, millions of robots are being used around the globe to perform various types of tasks, including materials-handling, spot-welding, arc-welding, and routing. A robot may simply be described as a mechanism guided by automatic controls. The idea of the functional robot may be traced back to the writings of Aristotle (4th century BC), the teacher of Alexander the Great, in which he wrote "If every instrument could accomplish its own work, obeying or anticipating the will of others ...." [1].

The word "robot" is derived from the Czechoslovakian language, in which it means "worker" [2]. The first commercial robot was manufactured by Planet Corporation, in 1959 [3]. As robots use mechanical, pneumatic, hydraulic, electrical, and electronic parts, their reliability-related problems are very challenging because of the sources of failures are many and varied. Although there is no clear-cut definitive point for the beginnings of the field of robot reliability, articles by J.F. Engelberger and K.M. Haugan, in 1974, could be regarded as its starting point [4,5]. In 1987, an article presented a comprehensive list of publications that were directly or indirectly related to robot reliability [6] and, in 1991, a book titled *Robot Reliability and Safety* covered the topic of robot reliability in considerable depth [7].

Many articles, directly or indirectly, in the 1970s and early 1980s covered the topic of robot safety [6]; in 1985 and 1986, two standards titled "An Interpretation of the Technical Guidance on Safety Standards in the Use, etc. of Industrial Robots" and "Industrial Robots and Robot Systems—Safety Requirements" were published by the Japanese Industrial Safety and Health Association [8], and the American National Standards Institute [9], respectively. Also, in 1985, an edited book titled *Robot Safety* was published [10].

Needless to say, since the 1980s, a large number of publications have appeared that are directly or indirectly related to robot-system reliability and safety. A list of over 370 such publications is provided in the Appendix.

1.2 Robot System Reliability/Safety-Related Facts, Figures, and Examples

Some of the facts, figures, and examples directly or indirectly concerned with robot-system reliability/safety are as follows:

- As per [11], some studies conducted in Japan indicate that more than 50% of working accidents with robots can be attributed to faults in the control system's electronic circuits.
- As per [11], the Japanese studies also indicate that "human error" was responsible for less than 20% of working accidents with robots.
- A study reported that 12–17% of the accidents in industries using advanced manufacturing technology were related to automated production equipment [12,13].
- The first robot-related fatal accidents occurred in Japan in 1978 and in the United States in 1984 [14,15].
- A material-handling robot was operating in its automatic mode and a worker violated all safety devices in entering the robot work-cell. The worker got trapped between the robot and a post anchored to the floor, was consequently injured and died a few days later [7,16,17].
- For the period 1978–1987, there were 10 robot-related fatal accidents in Japan [15].
- A worker entered the robot cell to clean its sensors and was killed because he/she ignored did not observe stated lockout procedures [7,16–20].
- As per [12], during the period 1978–1984, there were at least five robot-related fatal accidents: one in the United States and four in Japan.
- A maintenance worker climbed over a safety fence without turning off robot power and performed tasks in the robot work area while it was temporarily stopped. When the robot recommenced operation, it pushed the maintenance worker into a grinding machine and the worker died subsequently [7,16,17].
- In 1987, a study of 32 robot-related accidents that occurred in four countries—Japan, United States, West Germany, and Sweden— reported that line workers were at the greatest risk of injury followed by maintenance workers [14].
- A worker stepped between a robot and the machine (a planer) it was servicing, and turned off the power to the circuit transmitting/

activating signal from the planer to the robot. After accomplishing the necessary task, the worker switched on the power to the same circuit while still in the robot's work area. When the robot recommenced its operation, it crushed the worker to death against the planer [7,16,17].

- A robot operator went to check/fix a robot failure without locking out the robot. During the process, the operator activated the robot and the arm of the robot crushed the operator against a part being transported on a conveyor and the operator died [7,16–20].

- A worker holding the robot controls in his/her hand activated the robot while bending over the wheel to check the settings. The robot pinned the worker against the wheel and crushed him/her to death [7,16–20].

- A worker turned on a welder-robot while another worker was still in the robot's work area. The robot pushed the worker in its work area into the positioning fixture to death [7,16–20].

- A worker climbed onto a conveyor belt in motion for recovering a faulty part when the robot servicing the line was stopped on a program point temporarily. The robot crushed the worker to death when the belt resumed its operation [7,16–20].

- A worker entered the caged robotic palletizer cell while the robotic palletizer was still running. The worker's torso was crushed by the arms of the robotic palletizer as it tried to pick up boxes on the roller conveyor and the worker died [7,16–20].

- A worker was freeing a jam at the end of an oven and became caught between the conveyor belt of the oven and the robotic arm and got killed [7,16–20].

- A worker was troubleshooting a robotic arm used for removing CD jewel cases from an injection molding machine, when the arm cycled and struck the worker, who died two weeks later [7,16–20].

1.3 Terms and Definitions

This section presents some useful terms and definitions directly or indirectly concerned with the reliability and safety of robot systems [7,9,21–29].

- *Robot.* This is an automatic, reprogrammable, position-controlled, multifunctional manipulator, consisting of several axes designed for moving materials, components, tools, or specialized devices through variable programmed motions to perform various tasks.

- *Reliability.* This is the probability that an item/system will perform its stated mission satisfactorily for the specified time period when used according to the specified conditions.
- *Safety.* This is conservation of human life and the prevention of damage to items as per mission-stated requirements.
- *End-effector.* This is a gripper, actuator, or driven mechanical device designed for attachment to the robot wrist (i.e., the end of a manipulator) by which items can be acted upon.
- *Pendant.* This is a portable control device, including teaching pendants, that permits a human to control the robot from within its (i.e., the robot's) work area or zone.
- *Barrier.* This is a physical means of separating humans from the prohibited robot work zone or area.
- *Gripper.* This is the grasping hand of the robot which manipulates items and tools to accomplish a specified task.
- *Error recovery.* This is the capability of intelligent robotic systems to reveal various types of errors and, through programming, to undertake corrective actions to overcome the impending problem and complete the specified process.
- *Erratic robot.* A robot that has moved appreciably off its stated path.
- *Hazardous motion.* This is a sudden robot motion that may result in injury.
- *Fail-safe.* This is the failure of a robot/robot part, without endangering humans or damage to equipment or plant facilities.
- *Robot mean time to failure.* This is the average time that a robot will operate before failure.
- *Awareness barrier.* This is a device or an attachment that, by physical and visual means, alerts humans regarding potential or existing hazards.
- *Robot mean time to repair.* This is the average time that a robot is expected to be out of action after failure.
- *Safeguard.* This is a barrier device, guard, or procedure developed for human protection.
- *Downtime.* This is the time period during which the item/system/robot is not in a condition to carry out its stated mission.
- *Mission time.* This is the element of uptime that is required to perform a stated mission profile.
- *Random failure.* This is any failure whose occurrence cannot be predicted.
- *Redundancy.* The existence of more than one means to perform a specified function.

- *Accident.* This is an event that involves damage to a certain system/item/robot that suddenly disrupts the potential or current system/item/robot output.
- *Continuous task.* This is a task/job that involves some kind of tracking activity (i.e., monitoring a changing condition).

1.4 Useful Sources for Obtaining Information on Reliability and Safety of Robot Systems

This section lists journals, conference proceedings, books, technical reports, data sources, standards, and organizations directly or indirectly considered quite useful to obtain information on the reliability and safety of robot systems.

1.4.1 Journals

- *Robotica*
- *Robotics*
- *Journal of Occupational Accidents*
- *Safety Science*
- *Accident Analysis and Prevention*
- *International Journal of Human Factors Engineering*
- *Journal of Robotic Systems*
- *Robotics World*
- *The Industrial Robot*
- *Soviet Engineering Research*
- *International Journal of Robotics and Automation*
- *Journal of Safety Research*
- *Journal of Mechanical Design*
- *Professional Safety*
- *Reliability Engineering and System Safety*
- *IEEE Transactions on Reliability*
- *Robotics Engineering*
- *Microelectronics and Reliability*
- *Robotics and Autonomous Systems*
- *International Journal of Human Factors in Manufacturing*

- *Robotics Today*
- *International Journal of Robotics Research*
- *IEEE Transactions on Robotics and Automation*

1.4.2 Conference Proceedings

- *Proceedings of the IEEE International Conferences on Robotics and Automation*
- *Proceedings of the Annual Meetings of the Human Factors and Ergonomics Society*
- *Proceedings of the British Robot Association Annual Conferences*
- *Proceedings of the International Conferences on CAD/CAM, Robotics and Factories of the Future*
- *Proceedings of the IEEE International Conferences on Systems, Man, and Cybernetics*
- *Proceedings of the Annual Reliability and Maintainability Symposium*
- *Proceedings of the ANS Topical Meetings on Robotics and Remote Systems*
- *Proceedings of the IEEE International Symposium on Robot and Human Interactive Communication*

1.4.3 Books

- Bonney, M.C., Yong, Y.E., Editors, *Robot Safety*, Springer-Verlag, New York, 1985.
- Graham, J.H., Editor, *Safety, Reliability, and Human Factors*, Van Nostrand Reinhold, New York, 1991.
- Dhillon, B.S., *Robot Reliability and Safety*, Springer-Verlag, New York, 1991.
- Nof, S.Y., Editor, *Handbook of Industrial Robotics*, John Wiley and Sons, New York, 2007.
- Kurfess, T.R., Editor, *Robotics and Automation Handbook*, CRC Press, Boca Raton, Florida, 2005.
- Siciliano, B., Khatib, O., Editor, *Handbook of Robotics*, Springer-Verlag, Berlin, 2008.
- Dhillon, B.S., *Design Reliability: Fundamentals and Applications*, CRC Press, Boca Raton, Florida, 1999.
- Dhillon, B.S., *Engineering Safety: Fundamentals, Techniques, and Applications*, World Scientific Publishing, River Edge, New Jersey, 2003.
- Tver, D.F., Bolz, R.W., *Robotics Sourcebook and Dictionary*, Industrial Press, New York, 1983.

- Noro, K., Editor, *Occupational Health and Safety in Automation and Robotics*, Taylor & Francis, London, 1987.
- Shooman, M.L., *Probabilistic Reliability: An Engineering Approach*, McGraw Hill Book Company, New York, 1968.
- Handley, W., *Industrial Safety Handbook*, McGraw Hill Book Company, New York, 1969.
- Geotsch, D.L., *Occupational Safety and Health*, Prentice-Hall, Englewood Cliffs, New Jersey, 1996.
- Stephans, R.A., Talso, W.W., Editors, *System Safety Analysis Handbook*, System Safety Society, Irvine, California, 1993.
- Dhillon, B.S., *Reliability, Quality, and Safety for Engineers*, CRC Press, Boca Raton, Florida, 2005.
- Siegwart, R., Nourbakhsh, I.R., *Introduction to Autonomous Mobile Robots*, MIT Press, Cambridge, Massachusetts, 2011.

1.4.4 Technical Reports

- Hetzler, W.E., Hirsh, G.L., Machine operator crushed by robotic platform, *Nebraska Fatality Assessment and Control Evaluation (FACE) Investigation Report No. 99NE017*, The Nebraska Department of Labor, Omaha, Nebraska, October 25, 1999.
- Ulrich, K.T., Tuttle, T.T., Donoghue, J.P., Townsend, W.T., Intrinsically safer robots, *NASA Report No. NAS 10-12178*, Bararett Technology, Inc., Cambridge, Massachusetts, May 1995.
- Addison, J.H., Robotic safety systems and methods: Savannah River Site, *Report No. DPST-84-907 (DE 35-008261)*, December 1984, issued by E.E. du Pont de Nemonts and Co., Savannah River Laboratory, Aiken, South Carolina.
- National Institute of Occupational Safety and Health (NIOSH), Preventing the injury of workers by robots, *DHHS (NIOSH) Publication No. 85-103*, Morgantown, West Virginia, 1984.
- Japanese Ministry of Labor, Study on accidents involving robots, *Report No. PB 83239822*, Tokyo, 1982, available from the National Technical Information Service (NTIS), Springfield, Virginia.

1.4.5 Data Sources

- Reliability Analysis Center, Rome Air Development Center (RADC), Griffis Air Force Base, Rome, New York.
- Government/Industry Data Exchange Program (GIDEP), GIDEP Operations Center, US Department of the Navy, Corona, California.

- International Occupational Safety and Health Information Center Bureau, International du Travail, CH-1211, Geneva 22, Switzerland.
- Safety Research Information Service (SRIS), National Safety Council, 444 North Michigan Avenue, Chicago, Illinois.
- MIL-HANDBOOK 217, Rome Air Development Center, Griffis Air Force Base, Department of Defense, Rome, New York.
- Computer Accident/Incident Report System, System Safety Development Center, EG&G, PO Box 1625, Idaho Falls, Ohio.
- National Technical Information Service (NTIS), United States Department of Commerce, 5285 Port Royal Road, Springfield, Virginia.
- American National Standards Institute (ANSI), 11 W. 42nd St., New York, 10036.

1.4.6 Standards

- ANSI/RIA R15.06-2012, *American National Standard for Industrial Robots and Robot Systems-Safety Requirements*, American National Standards Institute, New York, 2012.
- *An Interpretation of the Technical Guidance on Safety Standards in the Use, etc., of Industrial Robots*, Japanese Industrial Safety and Health Association, Tokyo, 1985.
- RIA TR T15.106-2006, *Technical Report for Industrial Robots and Robot Systems-Safety Requirements Teaching Multiple Robots*, Robotic Industries Association, Ann Arbor, Michigan, 2006.
- MIL-STD-721, *Definitions of Terms for Reliability and Maintainability*, Department of Defense, Washington, DC.
- MIL-STD-882, *Systems Safety Program for System and Associated Subsystem and Equipment-Requirements*, Department of Defense, Washington, DC.
- MIL-STD-785, *Reliability Program for Systems and Equipment, Development and Production*, Department of Defense, Washington, DC.
- IEC 60950, *Safety of Information Technology Equipment*, International Electro-Technical Commission, Geneva, Switzerland, 1999.
- ISO 13482: 2014 (en), *Robots and Robotic Devices-Safety Requirements for Personal Care Robots*, International Organization for Standardization (ISO), Geneva 20, Switzerland, 2014.
- CAN/CAS-Z434-03 (R2013), *Industrial Robots and Robot Systems— General Safety Requirements*, Canadian Standards Association, Toronto, Canada, 2013.
- MIL-STD-2155, *Failure Reporting, Analysis, and Corrective Action (FRACAS)*, Department of Defense, Washington, DC.

- MIL-STD-756, *Reliability Modeling and Prediction*, Department of Defense, Washington, DC.
- MIL-STD-1629, *Procedures for Performing Failure Mode, Effects and Criticality Analysis*, Department of Defense, Washington, DC.

1.4.7 Organizations

- American Society of Safety Engineers, 1800 East Oakton St., Des Plaines, Illinois.
- British Safety Council, 62 Chancellors Road, London, U.K.
- Reliability Society, IEEE, PO Box 1331, Piscataway, New Jersey.
- Occupational Safety and Health Administration, U.S. Department of Labor, 200 Constitution Avenue, Washington, DC.
- The Robotics Society of America, PO Box 1205, Danville, California 94 526-1205.
- The International Federation of Robotics, Lyoner Str. 18, Frankfurt, Germany.
- Advanced Robot Technology Research Association, Kikai-shinko Bldg, 3-5-8 Shiba-Kohen, Minato-ku, Tokyo, Japan.
- American Society of Mechanical Engineers, 345 E. 47th Street, New York, New York 10017.
- Institute of Electrical and Electronics Engineers, Service Center, 445 Hoes Lane, Piscataway, New Jersey 08854-4150.
- Robotic Industries Association, 900 Victors Way, PO Box 3724, Ann Arbor, Michigan 48106.
- British Robot Association, BRA Aston Science Park, Love Lane, Birmingham, U.K.
- Japan Industrial Robot Association, 3-5-8 Shibakoen, Minato-Ku, Tokyo, Japan.

1.5 Scope of the Book

Robots are increasingly being used around the globe to perform various types of tasks. Some examples of such tasks are materials-handling, spot-welding, routing, and arc-welding. A robot has to be safe and reliable; today, reliability and safety of robot systems have become pressing issues because of the increasing number of accidental deaths and injuries; and cost.

Over the years, a large number of publications (i.e., journal and conference proceedings articles, technical reports, etc.) on robot system reliability and safety have appeared in the literature. At present, to the best of the author's knowledge, there is no book on the topic that covers recent developments in the area. Therefore, this book not only attempts to cover the reliability and safety of robot system within its framework, but also provides the latest developments in the area.

Finally, the main objective of this book is to provide professionals and others concerned with robot systems up-to-date information that could be useful for improving reliability and safety of such systems. This book will be useful to many individuals, including professionals working on reliability and safety in the robotics-related industrial sector, researchers, instructors, and graduate and senior undergraduate students in the area of robotics, and engineers-at-large.

1.6 Problems

1. Write an essay on reliability and safety of robot systems.
2. Define the following four terms:
   - Robot
   - End-effector
   - Safety
   - Reliability
3. List five important facts and figures concerning the reliability and safety of robot systems.
4. List at least six useful data sources for obtaining information related to the reliability and safety of robot systems.
5. What is the difference between the terms "awareness barrier" and "safeguard?"
6. List at least five books considered important to obtain information related to the reliability and safety of robot systems.
7. Define the following terms:
   - Robot mean time to failure
   - Accident
   - Hazardous motion
   - Error recovery
8. List at least six important organizations to obtain information concerning the reliability and safety of robot systems.
9. Discuss at least five important examples of robot-related accidents.
10. List at least six important standards directly or indirectly concerned with the reliability and safety of robot systems.

References

1. Heer, E., Robots in modern industry, in *Recent Advances in Robotics*, edited by G. Beni and S. Hackwood, John Wiley and Sons, New York, 1985, pp. 11–36.
2. Jablonowski, J., Posey, J.W., Robotics terminology, in *Handbook of Industrial Robotics*, edited by S.Y. Nof, John Wiley and Sons, New York, 1985, pp. 1271–1303.
3. Zeldman, M.I., *What Every Engineer Should Know About Robots*, Marcel Dekker, New York, 1984.
4. Engelberger, J.F., Three million hours of robot field experience, *Industrial Robot*, Vol. 1, No. 4, 1974, pp. 164–168.
5. Haugan, K.M., Reliability in industrial robots for spray gun applications, *Proceedings of 2nd Conference on Industrial Robot Technology*, University of Birmingham, U.K., 1974, pp. 1–5.
6. Dhillon, B.S., On robot reliability and safety: Bibliography, *Microelectronics and Reliability*, Vol. 27, No. 1, 1987, pp. 105–118.
7. Dhillon, B.S., *Robot Reliability and Safety*, Springer-Verlag, New York, 1991.
8. *An Interpretation of the Technical Guidance on Safety Standards in the Use, etc., of Industrial Robots*, Japanese Industrial Safety and Health Association, Tokyo, 1985.
9. ANSI/RIA R15.06-1986, *American National Standard for Industrial Robots and Robot Systems–Safety Requirements*, American National Standards Institute, New York, 1986.
10. Bonney, M.C., Yong, Y.F., Editors, *Robot Safety*, Springer-Verlag, New York, and IFS (Publications), Bedford, England, 1985.
11. Retsch, T., Schmitter, G., Marty, A., Safety principles for industrial robots, in *Encyclopedia of Occupational Health and Safety*, Vol. II, edited by J.M. Stellman, International Labor Organization, Geneva, Switzerland, 2011, pp. 58.56–58.58.
12. Backtrom, T., Dooes, M., A comparative study of occupational accidents in industries with advanced manufacturing technology, *International Journal of Human Factors in Manufacturing*, Vol. 5, 1995, pp. 267–282.
13. Clark, D.R., Lehto, M.R., Reliability, maintenance, and safety of robots, in *Handbook of Industrial Robotics*, edited by S.Y. Nof, John Wiley and Sons, New York, 1999, pp. 717–753.
14. Sanderson, L.M., Collins, J.N., McGlothlin, J.D., Robot-related fatality involving a U.S manufacturing plant employee: Case report and recommendations, *Journal of Occupational Accidents*, Vol. 8, 1986, pp. 13–23.
15. Nagamachi, M., Ten fatal accidents de to robots in Japan, in *Ergonomics of Hybrid Automated Systems*, edited by W. Karwowski et al., Elsevier, Amsterdam, 1988, pp. 391–396.
16. Japanese Ministry of Labor, *Study on Accidents Involving Industrial Robots, Report No. PB 83239822*, Tokyo, 1982, available from the National Technical Information Service (NTIS), Springfield, Virginia.
17. Nicolaisen, P., Safety problems related to robots, *Robotics*, Vol. 3, 1987, pp. 205–211.
18. Altamuro, V.M., Working safely with the iron collar worker, *National Safety News*, July 1983, pp. 38–40.
19. Occupational Safety and Health Administration (OSHA), *Report No. 0552652*, Washington, DC, October 10, 2006.

20. Lauch, K.E., New Standards for Industrial Robot Safety, *CIM Review*, Spring 1986, pp. 60–68.
21. Fisher, E.L., ed., Glossary of robotics terminology, in *Robotics*, Industrial Engineering and Management Press, Institute of Industrial Engineers, Atlanta, Georgia, 1983, pp. 231–253.
22. Tver, D.F., Boltz, R.W., *Robotics Source Book and Dictionary*, Industrial Press, New York, 1983.
23. Susnjara, K.A., *A Manager's Guide to Industrial Robots*, Corinthian Press, Shaker Heights, Ohio, 1982.
24. US Department of Defense, *MIL-STD-721B, Definitions of Effectiveness Terms for Reliability, Maintainability, Human Factors, and Safety*, Washington, DC, August 1966.
25. American Society of Safety Engineers, *Dictionary of Terms Used in the Safety Profession*, 3rd Edition, Des Plaines, Illinois, 1988.
26. McKenna, T., Oliverson, R., *Glossary of Reliability and Maintenance Terms*, Gulf Publishing Company, Houston, Texas, 1997.
27. Meulen, M.V.D., *Definitions for Hardware and Software Safety Engineers*, Springer-Verlag, London, 2000.
28. Dhillon, B.S., *Engineering Safety: Fundamentals, Techniques, and Applications*, World Scientific Publishing, River Edge, New Jersey, 2003.
29. Dhillon, B.S., *Applied Reliability and Quality: Fundamentals, Methods, and Procedures*, Springer-Verlag, London, 2007.

2

Basic Mathematical Concepts

2.1 Introduction

As in the development of other areas of science and technology, mathematics has played an important role in the development of the reliability and safety fields of the robot systems also. The history of mathematics may be traced back to the development of our currently used number symbols, sometimes referred as the "Hindu-Arabic numeral system" in the published literature [1]. Among the early evidences of the use of these number symbols are notches found on stone columns erected by the Scythian Emperor of India named Asoka, in around 250 BC [1].

The earliest reference to the concept of probability may be traced back to a gambler's manual written by Girolamo Cardano (1501–1576) [2]. However, Blaise Pascal (1623–1662) and Pierre Fermat (1601–1665) were the first two individuals who solved independently and correctly the problem of dividing the winnings in a game of chance [1,2]. Boolean algebra, which plays a pivotal role in modern probability theory, is named after an English mathematician George Boole (1815–1864), who published in 1847 a pamphlet titled "The Mathematical Analysis of Logic: Being an Essay towards a Calculus of Deductive Reasoning" [1–3].

Laplace transforms, frequently used in the area of reliability to find solutions to first-order differential equations, were developed by a French mathematician named Pierre-Simon Laplace (1749–1827). A more detailed history of mathematics and probability is available in [1,2]. This chapter presents basic mathematical concepts that will be useful to understand subsequent chapters of this book.

2.2 Arithmetic Mean and Mean Deviation

A given set of data related to reliability/safety of robot systems is useful only if it is analyzed properly. More clearly, there are certain characteristics of the data that are useful to describe the nature of a given data-set, thus making better associated decisions. Thus, this section presents two statistical measures considered useful to study reliability and safety-related data for robot systems [4–6].

2.2.1 Arithmetic Mean

This is expressed by

$$m = \frac{\sum_{j=1}^{n} y_j}{n} \tag{2.1}$$

where
   m is the mean value (i.e., arithmetic mean),
   y_j is the data value j, for $j = 1,2,3,\ldots,n$ and
   n is the number of data values.

It is to be noted that the arithmetic mean is generally simply called "mean."

> **EXAMPLE 2.1**
> Assume that the quality control department of a robot system manufacturing company inspected five identical robot systems and discovered 1, 5, 3, 4, and 2 defects in each respective robot system. Calculate the average number of defects (i.e., arithmetic mean) per robot system.
>   By inserting the given data values into Equation 2.1, we obtain:
>
> $$m = \frac{1+5+3+4+2}{5} = 3 \text{ defects per robot system}$$
>
>   Thus, the average number of defects per robot system is three. In other words, the arithmetic mean of the given dataset is 3.

2.2.2 Mean Deviation

Mean deviation is a measure of dispersion whose value indicates the degree to which a given set of data tends to spread about a mean value. Mean deviation is expressed by

$$\text{MD} = \frac{\sum_{j=1}^{k} \left| y_j - m \right|}{k} \tag{2.2}$$

where
   k is the number of data values,
   MD is the mean deviation,
   m is the mean value of the given dataset,

y_j is the data value j, for $j = 1,2,3,...,k$ and

$|y_j - m|$ is the absolute value of the deviation of y_j from m.

EXAMPLE 2.2

Calculate the mean deviation of the dataset given in Example 2.1.

In Example 2.1, the calculated mean value (i.e., arithmetic mean) of the given dataset is three defects per robot system. Thus, using this calculated value and the given data values in Equation 2.2, we obtain:

$$MD = \frac{|1-3| + |5-3| + |3-3| + |4-3| + |2-3|}{5}$$

$$= \frac{[2+2+0+1+1]}{5}$$

$$= 1.2$$

Thus, the mean deviation of the dataset in Example 2.1 is 1.2.

2.3 Boolean Algebra Laws

Boolean algebra, named after its founder, George Boole (1815–1864), is used to a degree in the studies of reliability and safety-related in robot systems. Some of its laws that are considered useful to understand subsequent chapters of this book are presented below [3–5,7,8].

$$C.D = D.C \tag{2.3}$$

where

C is an arbitrary set or event,

D is an arbitrary set or event and

Dot(.) denotes the intersection of sets.

It is to be noted that sometimes Equation 2.3 is written without the dot (e.g., CD), but it still conveys the same meaning.

$$C + D = D + C \tag{2.4}$$

where

+ is the union of sets or events.

$$C + C = C \tag{2.5}$$

$$CC = C \tag{2.6}$$

$$D(D + C) = D \tag{2.7}$$

$$C + CD = C \tag{2.8}$$

$$C(D + E) = CD + CE \tag{2.9}$$

where
   E is an arbitrary set or event.

$$(C + D)(C + E) = C + D\,E \tag{2.10}$$

$$(C + D) + E = C + (D + E) \tag{2.11}$$

$$(CD)E = C(DE) \tag{2.12}$$

It is to be noted that in the published literature, Equations 2.3 and 2.4 are referred to as commutative law, Equations 2.5 and 2.6 as idempotent law, Equations 2.7 and 2.8 as absorption law, Equations 2.9 and 2.10 as distributive law, and Equations 2.11 and 2.12 as associative law [9].

2.4 Probability Definition and Properties

The probability is defined as follows [9,10]:

$$P(Z) = \lim_{m \to \infty} \left(\frac{M}{m} \right) \tag{2.13}$$

where
   $P(Z)$ is the occurrence probability of event Z and
   M is the number of times event Z occurs in the m repeated experiments.

Some of the basic properties of probability are as follows [8–10]:

• The probability of the occurrence of an event, say event Z, is

$$0 \leq P(Z) \leq 1 \tag{2.14}$$

• The probability of the occurrence and nonoccurrence of an event, say event Z, is always:

$$P(Z) + P(\bar{Z}) = 1 \tag{2.15}$$

where
 $P(Z)$ is the probability of the occurrence of event Z.
 $P(\bar{Z})$ is the probability of the nonoccurrence of event Z.

- The probability of an intersection of m-independent events is

$$P(Z_1 Z_2 Z_3 \ldots Z_m) = P(Z_1)P(Z_2)P(Z_3)\,P(Z_m) \tag{2.16}$$

where
 $P(Z_j)$ is the probability of the occurrence of event Z_j, for $j = 1,2,3,\ldots,m$.

- The probability of the union of m-independent events is

$$P(Z_1 + Z_2 + \cdots + Z_m) = 1 - \prod_{j=1}^{m}(1 - P(Z_j)) \tag{2.17}$$

- The probability of the union of m mutually exclusive events is

$$P(Z_1 + Z_2 + \cdots + Z_m) = \sum_{j=1}^{m} P(Z_j) \tag{2.18}$$

EXAMPLE 2.3

Assume that a robot system is composed of two critical subsystems Z_1 and Z_2. The failure of either subsystem can, directly or indirectly, lead to a serious accident. The probability of failure of subsystems Z_1 and Z_2 is 0.01 and 0.02, respectively.

Calculate the probability of the occurrence of a serious accident in a robot system if both of these subsystems fail independently.

By inserting the specified data values into Equation 2.17, we obtain:

$$\begin{aligned}
P(Z_1 + Z_2) &= 1 - \prod_{j=1}^{2}(1 - P(Z_j)) \\
&= P(Z_1) + P(Z_2) - P(Z_1)P(Z_2) \\
&= 0.01 + 0.02 - (0.01)(0.02) \\
&= 0.0298
\end{aligned}$$

Thus, the occurrence probability of a serious robot system accident is 0.0298.

2.5 Probability Distribution-Related Definitions

This section presents three probability distribution-related definitions considered useful to perform various types of reliability and safety studies concerned with robot systems.

2.5.1 Cumulative Distribution Function

For a continuous random variable, the cumulative distribution function is
expressed by [9,10]

$$F(t) = \int_{-\infty}^{t} f(x)\, dx \tag{2.19}$$

where
  x is a continuous random variable,
  $f(x)$ is the probability density function, and
  $F(t)$ is the cumulative distribution function.

For $t = \infty$, Equation 2.19 becomes

$$F(\infty) = \int_{-\infty}^{\infty} f(x)\, dx$$

$$= 1 \tag{2.20}$$

It means that the total area under the probability density curve is equal to
unity.

Normally, in reliability and safety studies of robot systems, Equation 2.19
is simply written as

$$F(t) = \int_{0}^{t} f(x)\, dx \tag{2.21}$$

EXAMPLE 2.4

Assume that the probability (i.e., failure) density function of a robot system is

$$f(t) = \theta e^{-\theta t}, \quad \text{for } t \geq 0, \theta > 0 \tag{2.22}$$

where
  t is a continuous random variable (i.e., time),
  θ is robot system failure rate and

f(t) is the probability density function (generally, in the area of reliability-engineering, it is called the failure density function).

Obtain an expression for the robot system cumulative distribution function with the aid of Equation 2.22.

By inserting Equation 2.22 into Equation 2.21, we get

$$F(t) = \int_0^t \theta e^{-\theta t} dt$$
$$= 1 - e^{-\theta t} \qquad (2.23)$$

Thus, Equation 2.23 is the expression for the robot system cumulative distribution function.

2.5.2 Probability Density Function

For a continuous random variable, the probability density function is defined by [10]

$$f(t) = \frac{dF(t)}{dt} \qquad (2.24)$$

where

f(t) is the probability density function and

F(t) is the cumulative distribution function.

EXAMPLE 2.5

Prove by using Equation 2.23 that Equation 2.22 is the probability density function.

By substituting Equation 2.23 into Equation 2.24, we get

$$f(t) = \frac{d(1 - e^{-\theta t})}{dt}$$
$$= \theta e^{-\theta t} \qquad (2.25)$$

Equations 2.25 and 2.22 are identical.

2.5.3 Expected Value

The expected value of a continuous random variable is defined by [10]

$$E(t) = \int_{-\infty}^{\infty} t f(t) dt \qquad (2.26)$$

where

E(t) is the expected value (i.e., mean value) of the continuous random variable *t*.

EXAMPLE 2.6

Find the mean value (i.e., expected value) of the probability (failure) density function expressed by Equation 2.22.

By substituting Equation 2.22 into Equation 2.26, we get

$$E(t) = \int_0^\infty t\theta e^{-\theta t}dt$$

$$= \left[-te^{-\theta t} \right]_0^\infty - \left[-\frac{e^{-\theta t}}{\theta} \right]_0^\infty$$

$$= \frac{1}{\theta} \tag{2.27}$$

Thus, the mean value (i.e., expected value) of the probability (failure) density function expressed by Equation 2.22 is given by Equation 2.27.

2.6 Probability Distributions

Although there are a large number of probability or statistical distributions in published literature, this section presents just four such distributions considered useful to perform various types of robot system reliability and safety studies [11–13].

2.6.1 Exponential Distribution

This is one of the simplest continuous random variable distributions often used in the industrial sector, particularly to perform reliability-related studies. Its probability density function is defined by [9,14]

$$f(t) = \theta e^{-\theta t}, \quad \text{for } \theta > 0, t \geq 0 \tag{2.28}$$

where
t is the time (i.e., a continuous random variable),
$f(t)$ is the probability density function, and
θ is the distribution parameter.

By substituting Equation 2.28 into Equation 2.21, we obtain the following expression for the cumulative distribution function:

$$F(t) = 1 - e^{-\theta t} \tag{2.29}$$

Using Equations 2.26 and 2.28, we get the following expression for the distribution expected value (i.e., mean value):

$$E(t) = \frac{1}{\theta} \tag{2.30}$$

EXAMPLE 2.7

Assume that the mean time to failure of a robot system is 2000 h. Calculate the robot system's probability of failure during a 1000-h mission by using Equations 2.29 and 2.30.

By substituting the given data value into Equation 2.30, we get

$$\theta = \frac{1}{2000} = 0.0005 \text{ failures per hour}$$

By inserting the calculated and the given data values into Equation 2.29, we obtain:

$$F(100) = 1 - e^{-(0.0005)(1000)}$$
$$= 0.3935$$

Thus, the robot system's probability of failure during the 1000-h mission is 0.3935.

2.6.2 Rayleigh Distribution

This continuous random variable probability distribution is named after its founder, John Raleigh (1842–1919) [1]. The probability density function of the distribution is defined by

$$f(t) = \left(\frac{1}{\gamma^2} \right) t e^{-(t/\gamma)^2}, \quad \text{for } \gamma > 0, t \geq 0 \tag{2.31}$$

where
γ is the distribution parameter.

By inserting Equation 2.31 into Equation 2.21, we get the following equation for the cumulative distribution function:

$$F(t) = 1 - e^{-(t/\gamma)^2} \tag{2.32}$$

By substituting Equation 2.31 into Equation 2.26, we obtain the following equation for the distribution expected value:

$$E(t) = \gamma \Gamma \left(\frac{3}{2} \right) \tag{2.33}$$

where
Γ(.) is the gamma function and is expressed by

$$\Gamma(m) = \int_0^\infty t^{m-1} e^{-t} dt, \quad \text{for } m > 0 \tag{2.34}$$

2.6.3 Weibull Distribution

This continuous random variable distribution was developed by Walliodi Weibull, a Swedish professor in mechanical engineering in the early 1950s [15]. The distribution probability density function is defined by

$$f(t) = \frac{ct^{c-1}}{\gamma^c} e^{-(t/\gamma)^c}, \quad \text{for } \gamma > 0, c > 0, t \geq 0 \tag{2.35}$$

where
c and γ are the distribution shape and scale parameters, respectively.

By substituting Equation 2.35 into Equation 2.21, we get the following equation for the cumulative distribution function:

$$F(t) = 1 - e^{-(t/\gamma)^c} \tag{2.36}$$

By inserting Equation 2.35 into Equation 2.26, we get the following equation for the distribution expected value:

$$E(t) = \gamma \Gamma \left(1 + \frac{1}{c} \right) \tag{2.37}$$

It is to be noted that for $c = 2$ and $c = 1$, the Rayleigh and exponential distributions are the special cases of this distribution, respectively.

2.6.4 Bathtub Hazard Rate Curve Distribution

This is another continuous random variable distribution and it was developed in 1981 [16]. In the published literature by other authors around the world, it is generally referred to as the Dhillon distribution/law/model [17–36]. The distribution can represent bathtub-shape, increasing, and decreasing hazard rates.

The probability density function of the distribution is defined by [16]

$$f(t) = c\gamma (\gamma t)^{c-1} e^{-\left\{ e^{(\gamma t)^c} - (\gamma t)^c - 1 \right\}}, \quad \text{for } t \geq 0, \gamma > 0, c > 0, \tag{2.38}$$

where
c and γ are the distribution shape and scale parameters, respectively.

By inserting Equation 2.38 into Equation 2.27, we get the following equation for the cumulative distribution function:

$$F(t) = 1 - e^{-\left\{ e^{(\gamma t)^c} - 1 \right\}} \tag{2.39}$$

It is to be noted that for $c = 0.5$, this probability distribution gives the bathtub-shaped hazard rate curve, and for $c = 1$ it gives the extreme value probability distribution. More clearly, the extreme value probability distribution is the special case of this probability distribution at $c = 1$.

2.7 Laplace Transform Definition, Common Laplace Transforms, and Final-Value Theorem Laplace Transform

This section presents various aspects of Laplace transforms considered useful to conduct robot system reliability and safety-related studies.

2.7.1 Laplace Transform Definition

The Laplace transform (named after a French mathematician, Pierre-Simon Laplace [1749–1827]) of a function, say $f(t)$, is defined by [1,37,38]

$$f(s) = \int_0^\infty f(t) e^{-st} dt \tag{2.40}$$

where
t is a variable,
s is the Laplace transform variable, and
f(s) is the Laplace transform of function f(t).

EXAMPLE 2.8

Obtain the Laplace transform of the following function:

$$f(t) = e^{-\lambda t} \tag{2.41}$$

where
   λ is a constant.

By substituting Equation 2.41 into Equation 2.40, we obtain:

$$f(s) = \int_{0}^{\infty} e^{-\lambda t} e^{-st} dt$$

$$= \frac{e^{-(s+\lambda)t}}{(s + \lambda)} \bigg|_{0}^{\infty}$$

$$= \frac{1}{s + \lambda} \tag{2.42}$$

Thus, Equation 2.42 is the Laplace transform of Equation 2.41.

2.7.2 Laplace Transforms of Common Functions

Laplace transforms of some commonly occurring functions in robot system reliability and safety-related analysis studies are presented in Table 2.1 [37–39].

TABLE 2.1

Laplace Transforms of Some Functions

No.	$f(t)$	$f(s)$
1	t^n, for $n = 0,1,2,3\ldots$	$\dfrac{n!}{s^{n+1}}$
2	K (a constant)	$\dfrac{K}{s}$
3	t	$\dfrac{1}{s^2}$
4	$e^{-\lambda t}$	$\dfrac{1}{s + \lambda}$
5	$te^{-\lambda t}$	$\dfrac{1}{(s + \lambda)^2}$
6	$t f(t)$	$-\dfrac{df(s)}{ds}$
7	$\dfrac{df(t)}{dt}$	$s f(s) - f(0)$
8	$\theta_1 f_1(t) + \theta_2 f_2(t)$	$\theta_1 f_1(s) + \theta_2 f_2(s)$

2.7.3 Final Value Theorem Laplace Transform

If the following limits exist, then the final value theorem may be stated as

$$\lim_{t \to \infty} f(t) = \lim_{s \to 0}[sf(s)] \tag{2.43}$$

EXAMPLE 2.9

Prove, by using the following equation, that the left-hand side of Equation 2.43 is equal to its right side:

$$f(t) = \frac{\theta_1}{(\theta_1 + \theta_2)} + \frac{\theta_2}{(\theta_1 + \theta_2)} e^{-(\theta_1 + \theta_2)t} \tag{2.44}$$

where
θ_1 and θ_2 are the constants.

By substituting Equation 2.44 into the left-hand side of Equation 2.43, we get

$$\lim_{t \to \infty}\left[\frac{\theta_1}{(\theta_1 + \theta_2)} + \frac{\theta_2}{(\theta_1 + \theta_2)} e^{-(\theta_1 + \theta_2)t}\right] = \frac{\theta_1}{\theta_1 + \theta_2} \tag{2.45}$$

With the aid of Table 2.1, we obtain the following Laplace transforms of Equation 2.44:

$$f(s) = \frac{\theta_1}{s(\theta_1 + \theta_2)} + \frac{\theta_2}{(\theta_1 + \theta_2)} \frac{1}{(s + \theta_1 + \theta_2)} \tag{2.46}$$

By inserting Equation 2.46 into the right-hand side of Equation 2.43, we obtain:

$$\lim_{s \to 0} s\left[\frac{\theta_1}{s(\theta_1 + \theta_2)} + \frac{\theta_2}{(\theta_1 + \theta_2)} \frac{1}{(s + \theta_1 + \theta_2)}\right] = \frac{\theta_1}{\theta_1 + \theta_2} \tag{2.47}$$

As the right-hand sides of Equations 2.45 and 2.47 are identical, it proves that the left-hand side of Equation 2.43 is equal to its right side.

2.8 Solving First-Order Differential Equations Using Laplace Transforms

Normally, Laplace transforms are used to find solutions to first-order linear differential equations in reliability and safety analysis-related studies of robot systems. The example presented below demonstrates the finding of solutions to a set of linear first-order differential equations, describing a robot system with respect to reliability, using Laplace transforms.

EXAMPLE 2.10

Assume that a robot system can be in any of the three states: operating normally, failed due to a hardware failure, or failed due to a human error. The following three first-order linear differential equations describe the robot system under consideration:

$$\frac{dP_0(t)}{dt} + (\lambda_1 + \lambda_2)P_0(t) = 0 \tag{2.48}$$

$$\frac{dP_1(t)}{dt} - \lambda_1 P_0(t) = 0 \tag{2.49}$$

$$\frac{dP_2(t)}{dt} - \lambda_2 P_0(t) = 0 \tag{2.50}$$

where
λ_1 is the robot system constant hardware failure rate,
λ_2 is the robot system constant human error rate, and
$P_j(t)$ is the probability that the robot system is in state j at time t, for $j = 0$ (operating normally), $j = 1$ (failed due to a hardware failure), and $j = 2$ (failed due to a human error),

At time $t = 0$, $P_0(0) = 1$, $P_1(0) = 0$, and $P_2(0) = 0$.
Solve differential Equations 2.48 through 2.50 with the aid of Laplace transforms.
With the aid of Table 2.1, differential Equations 2.48 through 2.50, and the given initial conditions, we get

$$sP_0(s) - 1 + (\lambda_1 + \lambda_2)P_0(s) = 0 \tag{2.51}$$

$$sP_1(s) - \lambda_1 P_0(s) = 0 \tag{2.52}$$

$$sP_2(s) - \lambda_2 P_0(s) = 0 \tag{2.53}$$

By solving Equations 2.51 through 2.53, we obtain:

$$P_0(s) = \frac{1}{(s + \lambda_1 + \lambda_2)} \tag{2.54}$$

$$P_1(s) = \frac{\lambda_1}{s(s + \lambda_1 + \lambda_2)} \tag{2.55}$$

$$P_2(s) = \frac{\lambda_2}{s(s + \lambda_1 + \lambda_2)} \tag{2.56}$$

By taking the inverse Laplace transforms of Equations 2.54 through 2.56, we get

$$P_0(t) = e^{-(\lambda_1 + \lambda_2)t} \tag{2.57}$$

$$P_1(t) = \frac{\lambda_1}{(\lambda_1 + \lambda_2)} \left[1 - e^{-(\lambda_1 + \lambda_2)t} \right] \tag{2.58}$$

$$P_2(t) = \frac{\lambda_2}{(\lambda_1 + \lambda_2)} \left[1 - e^{-(\lambda_1 + \lambda_2)t} \right] \tag{2.59}$$

Thus, Equations 2.57 through 2.59 are the solutions to differential Equations 2.48 through 2.50.

2.9 Problems

1. A robot system manufacturing company's quality control department inspected seven identical robot systems and found 4, 7, 8, 3, 2, 1, and 5 defects in each respective robot system. Calculate the average number of defects (i.e., the arithmetic mean) per robot system.
2. Calculate the mean deviation of the dataset given in Question 1.
3. What is commutative law?
4. Define probability.
5. What are the basic properties of probability?
6. Define expected value of a continuous random variable.
7. Define the following items:
   a. Probability density function; and
   b. Cumulative distribution function.
8. Write down the probability density function for the Rayleigh distribution.
9. Prove Equations 2.57 through 2.59 by using Equations 2.54 through 2.56.
10. What are the special case distributions of the Wiebull and bathtub hazard rate curve distributions?

References

1. Eves, H., *An Introduction to the History of Mathematics*, Holt, Rinehart and Winston, New York, 1976.
2. Owen, D.B., Editor, *On the History of Statistics and Probability*, Marcel Dekker, New York, 1976.
3. Lipschutz, S., *Set Theory*, McGraw Hill Book Company, New York, 1964.
4. Speigel, M.R., *Statistics*, McGraw Hill Book Company, New York, 1961.
5. Speigel, M.R., *Probability and Statistics*, McGraw Hill Book Company, New York, 1975.
6. Dhillon, B.S., *Reliability, Quality, and Safety for Engineers*, CRC Press, Boca Raton, Florida, 2004.
7. Fault Tree Handbook, Report No. NUREG-0492, U.S. Nuclear Regulatory Commission, Washington, DC, 1981.
8. Lipschutz, S., *Probability*, McGraw Hill Book Company, New York, 1965.
9. Dhillon, B.S., *Computer System Reliability: Safety and Usability*, CRC Press, Boca Raton, Florida, 2013.
10. Mann, N.R., Schafer, R.E., Singpurwalla, N.P., *Methods for Statistical Analysis of Reliability and Life Data*, John Wiley and Sons, New York, 1974.
11. Shooman, M.L., *Probabilistic Reliability: An Engineering Approach*, McGraw Hill Book Company, New York, 1968.
12. Patel, J.K., Kapadia, C.H., Owen, D.H., *Handbook of Statistical Distributions*, Marcel Dekker, New York, 1976.
13. Dhillon, B.S., *Reliability Engineering in Systems Design and Operation*, Van Nostrand Reinhold, New York, 1983.
14. Davis, D.J., Analysis of some failure data, *J. Am. Stat. Assoc.*, 1952, pp. 113–150.
15. Weibull, W., A statistical distribution function of wide applicability, *J. Appl. Mech.*, Vol. 18, 1951, pp. 293–297.
16. Dhillon, B.S., Life distributions, *IEEE Transactions on Reliability*, Vol. 30, 1981, pp. 457–460.
17. Baker, R.D., Non-parametric estimation of the renewal function, *Computers Operations Research*, Vol. 20, No. 2, 1993, pp. 167–178.
18. Cabana, A., Cabana, E.M., Goodness-of-fit to the exponential distribution, focused on Weibull alternatives, *Communications in Statistics-Simulation and Computation*, Vol. 34, 2005, pp. 711–723.
19. Grane, A., Fortiana, J., A directional test of exponentiality based on maximum correlations, *Metrika*, Vol. 73, 2011, pp. 255–274.
20. Henze, N., Meintnis, S.G., Recent and classical tests for exponentiality: A partial review with comparisons, *Metrika*, Vol. 61, 2005, pp. 29–45.
21. Jammalamadaka, S.R., Taufer, E., Testing exponentiality by comparing the empirical distribution function of the normalized spacings with that of the original data, *Journal of Nonparametric Statistics*, Vol. 15, No. 6, 2003, pp. 719–729.
22. Hollander, M., Laird, G., Song, K.S., Non-parametric interference for the proportionality function in the random censorship model, *Journal of Nonparametric Statistics*, Vol. 15, No. 2, 2003, pp. 151–169.

23. Jammalamadaka, S.R., Taufer, E., Use of mean residual life in testing departures from exponentiality, *Journal of Nonparametric Statistics*, Vol. 18, No. 3, 2006, pp. 277–292.

24. Kunitz, H., Pamme, H., The mixed gamma ageing model in life data analysis, *Statistical Papers*, Vol. 34, 1993, pp. 303–318.

25. Kunitz, H., A new class of bathtub-shaped hazard rates and its application in comparison of two test-statistics, *IEEE Transactions on Reliability*, Vol. 38, No. 3, 1989, pp. 351–354.

26. Meintanis, S.G., A class of tests for exponentiality based on a continuum of moment conditions, *Kybernetika*, Vol. 45, No. 6, 2009, pp. 946–959.

27. Morris, K., Szynal, D., Goodness-of-fit tests based on characterizations involving moments of order statistics, *International Journal of Pure and Applied Mathematics*, Vol. 38, No. 1, 2007, pp. 83–121.

28. Na, M.H., Spline hazard rate estimation using censored data, *Journal of KSIAM*, Vol. 3, No. 2, 1999, pp. 99–106.

29. Morris, K., Szynal, D., Some U-statistics in goodness-of-fit tests derived from characterizations via record values, *International Journal of Pure and Applied Mathematics*, Vol. 46, No. 4, 2008, pp. 339–414.

30. Nam, K.H., Park, D.H., Failure rate for Dhillon model, Proceedings of the Spring Conference of the Korean Statistical Society, 1997, pp. 114–118.

31. Nimoto, N., Zitikis, R., The Atkinson index, the Moran statistic, and testing exponentiality, *Journal of the Japan Statistical Society*, Vol. 38, No. 2, 2008, pp. 187–205.

32. Nam, K.H., Chang, S.J., Approximation of the renewal function for Hjorth model and Dhillon model, *Journal of the Korean Society for Quality Management*, Vol. 34, No. 1, 2006, pp. 34–39.

33. Noughabi, H.A., Arghami, N.R., Testing exponentiality based on characterizations of the exponential distribution, *Journal of Statistical Computation and Simulation*, Vol. 1, First, 2011, pp. 1–11.

34. Szynal, D., Goodness-of-fit tests derived from characterizations of continuous distributions, *Stability in Probability*, Banach Center Publications, Vol. 90, Institute of Mathematics, Polish Academy of Sciences, Warszawa, Poland, 2010, pp. 203–223.

35. Szynal, D., Wolynski, W., Goodness-of-fit tests for exponentiality and Rayleigh distribution, *International Journal of Pure and Applied Mathematics*, Vol. 78, No. 5, 2012, pp. 751–772.

36. Nam, K.H., Park, D.H., A study on trend changes for certain parametric families, *Journal of the Korean Society for Quality Management*, Vol. 23, No. 3, 1995, pp. 93–101.

37. Oberhettinger, F., Badii, L., *Tables of Laplace Transforms*, Springer-Verlag, Inc., New York, 1973.

38. Spiegel, M.R., *Laplace Transforms*, McGraw Hill Book Company, New York, 1965.

39. Nixon, F.E., *Handbook of Laplace Transformation: Fundamentals, Applications, Tables, and Examples*, Prentice-Hall, Inc., Englewood Cliffs, New Jersey, 1960.

3

Reliability and Safety Basics

3.1 Introduction

The history of the field of reliability goes back to the early years of the 1930s, when probability concepts were applied to problems in the area of electric power generation. However, the real beginning of the field is normally regarded as World War II, when German scientists applied basic reliability concepts for improving the reliability of their V1 and V2 rockets [1–4]. Today, the field of reliability has become a well-developed discipline and has branched into specialized areas such as power system reliability, human reliability, software reliability, and mechanical reliability [3,4].

The history of the field of safety may be traced back to 1868, when a patent for a barrier safeguard was awarded in the United States [5]. Twenty-five years later, in 1893, the US Congress passed the Railway Safety Act [5]. Today, the safety field has branched out into many specialized areas, including patient safety, workplace safety, and system safety [6].

This chapter presents various important aspects of reliability and safety basics considered useful to understand subsequent chapters of this book, as well as for the robotics sector at large.

3.2 Bathtub Hazard Rate Curve

Usually, this curve is used to describe the failure rate of engineering systems/items and is shown in Figure 3.1. The curve is called the bathtub hazard rate curve because it resembles the shape of a bathtub.

As shown in Figure 3.1, the curve is divided into three sections—the burn-in period, useful-life period, and wear-out period. During the burn-in period, the system/item hazard rate decreases with time t; some of the reasons for the occurrence of failures during this period are poor manufacturing methods and processes, substandard materials and workmanship, poor quality control, human error, and inadequate debugging [4,7]. Other terms

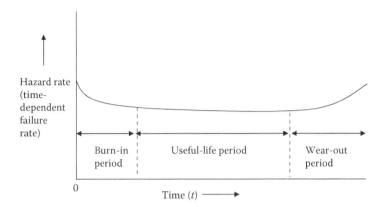

FIGURE 3.1
Bathtub hazard rate curve.

used in the published literature for this decreasing-hazard-rate region are "infant-mortality region," "break-in region," and "debugging region."

During the useful-life period, the hazard rate remains constant. Some of the causes for the occurrence of failures in this region are as follows [4,7]:

- Undetectable defects
- Low safety factors
- Human errors
- Higher random stress than expected
- Abuse
- Natural failures

Finally, during the wear-out period, the hazard rate increases with time *t* due to reasons such as wear from aging, poor maintenance practices, wrong overhaul practices, short designed-in life of the system/item under consideration, and wear due to friction, corrosion, and creep [4,7].

Mathematically, the following equation can be used to represent the bathtub hazard rate curve shown in Figure 3.1 [8]:

$$\lambda(t) = \theta a(\theta t)^{a-1} e^{(\theta t)^a} \tag{3.1}$$

where
    t is time,
    $\lambda(t)$ is hazard rate (time-dependent failure rate),
    θ is the scale parameter, and
    a is the shape parameter.

Equation 3.1 gives the shape of the bathtub hazard rate curve shown in Figure 3.1, at $a = 0.5$.

3.3 General Reliability-Related Formulas

A number of general formulas are often used to perform various types of reliability analysis. Four of these formulas that are based on the reliability function are presented below.

3.3.1 Probability (or Failure) Density Function

The probability (or failure) density function is expressed by [4]

$$f(t) = -\frac{dR(t)}{dt} \tag{3.2}$$

where
  $R(t)$ is the system/item reliability at time t and
  $f(t)$ is the probability (or failure) density function.

EXAMPLE 3.1

Assume that a robot system's reliability is expressed by

$$R_{rs}(t) = e^{-\lambda_{rs}t} \tag{3.3}$$

where
  $R_{rs}(t)$ is the robot system reliability at time t and
  λ_{rs} is the robot system constant failure rate.

Obtain an expression for the probability (or failure) density function of the robot system with the aid of Equation 3.2.

By substituting Equation 3.3 into Equation 3.2, we get

$$f(t) = -\frac{de^{-\lambda_{rs}t}}{\lambda_{rs}e^{-\lambda_{rs}t}} \tag{3.4}$$

Thus, Equation 3.4 is the expression for the probability (or failure) density function of the robot system.

3.3.2 Hazard Rate (or Time-Dependent Failure Rate) Function

This is expressed by

$$\lambda(t) = \frac{f(t)}{R(t)} \tag{3.5}$$

where

$\lambda(t)$ is the system/item hazard rate (or time-dependent failure rate).

By substituting Equation 3.2 into Equation 3.5, we obtain:

$$\lambda(t) = -\frac{1}{R(t)} \cdot \frac{dR(t)}{dt} \tag{3.6}$$

EXAMPLE 3.2

Obtain an expression for the hazard rate of the robot system with the aid of Equations 3.3 and 3.6 and comment on the final result. By inserting Equation 3.3 into Equation 3.6, we obtain:

$$\lambda_{rs}(t) = -\frac{1}{e^{-\lambda_{rs}t}} \cdot \frac{de^{-\lambda_{rs}t}}{dt}$$
$$= \lambda_{rs} \tag{3.7}$$

Thus, the hazard rate of the robot system is given by Equation 3.7. The right-hand side of this equation is not a function of time t. In other words, it is constant. Generally, it is referred to as the constant failure rate of a system/item (in this case, of the robot system) because it does not depend on time t.

3.3.3 General Reliability Function

This can be obtained by using Equation 3.6. Thus, using Equation 3.6, we get

$$-\lambda(t)dt = -\frac{1}{R(t)} \cdot dR(t) \tag{3.8}$$

By integrating both sides of Equation 3.8 over the time interval [0,t], we get

$$-\int_0^t \lambda(t)dt = \int_1^{R(t)} \frac{1}{R(t)} \cdot dR(t) \tag{3.9}$$

Since, at $t = 0$, $R(t) = 1$.

Evaluating the right-hand side of Equation 3.9 and then rearranging it yields

$$\ln R(t) = -\int_0^t \lambda(t)dt \tag{3.10}$$

Thus, from Equation 3.10, we get

$$R(t) = e^{-\int_0^t \lambda(t)dt} \tag{3.11}$$

Thus, Equation 3.11 is the general reliability function. This equation can be used to obtain the reliability function of a system/item when its times to failure follow any time-continuous probability distribution (e.g., exponential, Rayleigh, and Weibull).

EXAMPLE 3.3

Assume that the hazard rate of a robot system is expressed by Equation 3.1. Obtain an expression for the reliability function of the robot system with the aid of Equation 3.11.

By inserting Equation 3.1 into Equation 3.11, we get

$$R(t) = e^{-\int_0^t \{\theta a(\theta t)^{a-1} e^{(\theta t)^a}\}dt}$$

$$= e^{-\{e^{(\theta t)^a}-1\}} \tag{3.12}$$

Thus, Equation 3.12 is the expression for the reliability function of the robot system.

3.3.4 Mean Time to Failure

The mean time to failure of a system/item can be obtained by using any of the following three formulas [9,10]:

$$MTTF = \int_0^\infty R(t)dt \tag{3.13}$$

or

$$MTTF = \lim_{s \to 0} R(s) \tag{3.14}$$

or

$$MTTF = E(t) = \int_0^\infty t f(t) dt \qquad (3.15)$$

where
  $MTTF$ is the mean time to failure of a system/item,
  $E(t)$ is the expected value,
  s is the Laplace transform variable, and
  $R(s)$ is the Laplace transform of the reliability function $R(t)$.

EXAMPLE 3.4

Using Equation 3.3, prove that Equations 3.13 and 3.14 yield the same result for the robot system mean time to failure.
  By inserting Equation 3.3 into Equation 3.13, we get

$$MTTF_{rs} = \int_0^\infty e^{-\lambda_{rs}t} dt$$

$$= \frac{1}{\lambda_{rs}} \qquad (3.16)$$

where
  $MTTF_{rs}$ is the robot system mean time to failure.

By taking the Laplace transform of Equation 3.3, we obtain

$$R_{rs}(s) = \int_0^\infty e^{-st} e^{-\lambda_{rs}t} dt$$

$$= \frac{1}{(s + \lambda_{rs})} \qquad (3.17)$$

where
  $R_{rs}(s)$ is the Laplace transform of the robot system reliability function
    $R_{rs}(t)$.

By substituting Equation 3.17 into Equation 3.14, we get

$$MTTF_{rs} = \lim_{s \to 0} \frac{1}{(s + \lambda_{rs})}$$

$$= \frac{1}{\lambda_{rs}} \qquad (3.18)$$

As Equations 3.16 and 3.18 are identical, it proves that Equations 3.13 and 3.14 yield the same result for the robot system mean time to failure.

3.4 Reliability Configurations

In analyzing reliability, the analyst can encounter parts of a robot system forming a variety of different configurations. Thus, this section is concerned with the reliability-evaluation of such commonly occurring configurations or networks.

3.4.1 Series Network

This is the simplest network or configuration of reliability, and its block diagram is shown in Figure 3.2. The diagram represents an n-unit system, and each block in the diagram denotes a unit/part. If any one of the n units/parts fails, the series system/network/configuration fails. In other words, for the successful operation of the series system, all the n system units must operate normally.

The series system, reliability shown in Figure 3.2, reliability is expressed by

$$R_s = P(E_1E_2E_3 \ldots E_n) \tag{3.19}$$

where

E_j is the successful operation (i.e., success event) of unit j, for $j = 1,2,3,\ldots,n$.
$P(E_1E_2E_3\ldots E_n)$ is the occurrence probability of events $E_1E_2E_3\ldots E_n$.
R_s is the series system reliability.

For independently failing units, Equation 3.19 becomes

$$R_s = P(E_1)P(E_2)P(E_3)\ldots P(E_n) \tag{3.20}$$

where

$P(E_j)$ is the probability of occurrence of event E_j; for $j = 1,2,3,\ldots,n$.

If we let $R_j = P(E_j)$ for $j = 1,2,3,\ldots,n$, Equation 3.20 becomes

$$R_s = R_1R_2R_3 \ldots R_n$$

$$= \prod_{j=1}^{n} R_j \tag{3.21}$$

where

R_j is the unit j reliability, for $j = 1,2,3,\ldots,n$.

FIGURE 3.2
An n-unit series system/network/configuration.

For constant failure rate, λ_j, of unit j from Equation 3.11, we get

$$R_j(t) = e^{-\int_0^t \lambda_j dt}$$

$$= e^{-\lambda_j t} \tag{3.22}$$

where
$R_j(t)$ is the reliability of unit j at time t.

By substituting Equation 3.22 into Equation 3.21, we obtain

$$R_s(t) = e^{-\sum_{j=1}^n \lambda_j t} \tag{3.23}$$

where
$R_s(t)$ is the series system reliability at time t.

Substituting Equation 3.23 into Equation 3.13 yields the following expression for the series-system's mean time to failure:

$$MTTF_s = \int_0^\infty e^{-\sum_{j=1}^n \lambda_j t} dt$$

$$= \frac{1}{\sum_{j=1}^n \lambda_j} \tag{3.24}$$

where
$MTTF_s$ is the series-system's mean time to failure.

By inserting Equation 3.23 into Equation 3.6, we get the following expression for the series-system's hazard rate:

$$\lambda_s(t) = -\frac{1}{e^{-\sum_{j=1}^n \lambda_j t}} \left[-\sum_{j=1}^n \lambda_j \right] e^{-\sum_{j=1}^n \lambda_j t}$$

$$= \sum_{j=1}^n \lambda_j \tag{3.25}$$

where
$\lambda_s(t)$ is the series-system's hazard rate.

It is to be noted that the right-hand side of Equation 3.25 is independent of time t. Thus, the left-hand side of this equation is simply λ_s, the failure rate of the series system. It means that whenever we add up failure rates of independent items/units/parts, we automatically assume that these items/units/parts form a series network or configuration, a worst-case design scenario in regard to reliability.

EXAMPLE 3.5

Assume that a robot system is composed of four independent and identical subsystems and the constant failure rate of each subsystem is 0.0001 failures per hour. All four subsystems must operate normally for the robot system to function successfully.

Calculate the robot system's reliability for a 50-h mission, mean time to failure, and failure rate.

By inserting the given data values into Equation 3.23, we obtain:

$$R_s(50) = e^{-((0.0001+0.0001+0.0001+0.0001)(50))}$$
$$= 0.9802$$

Substituting the given data values into Equation 3.24 yields:

$$MTTF_s = \frac{1}{(0.0001 + 0.0001 + 0.0001 + 0.0001)}$$
$$= 2500 \text{ h}$$

By substituting the given data values into Equation 3.25, we get

$$\lambda_s = (0.0001 + 0.0001 + 0.0001 + 0.0001)$$
$$= 0.0004 \text{ failures per hour}$$

Thus, the robot system's reliability, mean time to failure, and failure rate are 0.9802, 2500 h, and 0.0004 failures per hour, respectively.

3.4.2 Parallel Network

This network represents a system with n units/items operating simultaneously. For the successful operation of the system, at least one of the n units must operate normally. The block diagram of an n-unit parallel system is shown in Figure 3.3; each block in the diagram represents a unit.

The probability of failure of the parallel system shown in Figure 3.3 is expressed by

$$F_p = P(\bar{y}_1\bar{y}_2\bar{y}_3\ldots\bar{y}_n) \tag{3.26}$$

where
F_p is the probability of failure of the parallel system.
$\bar{y}_j$ is the failure (i.e., failure event) of unit j, for $j = 1,2,3,\ldots,n$.
$P(\bar{y}_1\bar{y}_2\bar{y}_3\ldots\bar{y}_n)$ is the probability of occurrence of events $\bar{y}_1, \bar{y}_2, \bar{y}_3,\ldots,$ and $\bar{y}_n$.

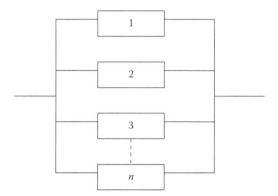

FIGURE 3.3
Block diagram of a parallel system or network with n units.

For independently failing parallel units, Equation 3.26 is written as

$$F_p = P(\bar{y}_1)P(\bar{y}_2)P(\bar{y}_3)\dots P(\bar{y}_n) \tag{3.27}$$

where
$P(\bar{y}_j)$ is the probability of occurrence of failure event $\bar{y}_j$ for $j = 1,2,3,\dots,n$.

If we let $F_j = P(\bar{y}_j)$, for $j = 1,2,3,\dots,n$, then Equation 3.27 becomes

$$F_p = \prod_{j=1}^{n} F_j \tag{3.28}$$

where
F_j is the unit j failure probability, for $j = 1,2,3,\dots,n$.

By subtracting Equation 3.28 from unity, we get

$$R_p = 1 - \prod_{j=1}^{n} F_j \tag{3.29}$$

where
R_p is the parallel system/network reliability.

For constant failure rate λ_j of unit j, subtracting Equation 3.22 from unity and then inserting it into Equation 3.29 yields:

$$R_p(t) = 1 - \prod_{j=1}^{n} (1 - e^{\lambda_j t}) \tag{3.30}$$

where
  $R_p(t)$ is the parallel system/network reliability at time t.

For identical units, Equation 3.30 becomes

$$R_p(t) = 1 - (1 - e^{-\lambda t})^n \tag{3.31}$$

where
  λ is the unit constant failure rate.

By substituting Equation 3.31 into Equation 3.13, we get the following expression for the parallel system or network mean time to failure:

$$MTTF_p = \int_0^\infty \left[1 - (1 - e^{-\lambda t})^n \right] dt$$

$$= \frac{1}{\lambda} \sum_{j=1}^n \frac{1}{j} \tag{3.32}$$

where
  $MTTF_p$ is the identical unit parallel system/network mean time to failure.

EXAMPLE 3.6

Assume that a robot system is composed of three identical, active, and independent units. At least one of the units must function normally for the robot system to operate successfully. The failure rate of a unit is 0.0002 failures per hour.

Calculate the robot system's reliability for a 300-h mission and mean time to failure.

By inserting the given data values into Equation 3.31, we obtain:

$$R_p(300) = 1 - [1 - e^{(0.0002)(300)}]^3$$
$$= 0.9998$$

Substituting the specified data values into Equation 3.32 yields:

$$MTTF_p = \frac{1}{(0.0002)} \left[1 + \frac{1}{2} + \frac{1}{3} \right]$$
$$= 9166.67 \text{ h}$$

Thus, the robot system's reliability and mean time to failure are 0.9998 and 9166.67 h, respectively.

3.4.3 *k*-out-of-*n* Network

In this case, the network/system is composed of n active units, and at least k units out of n active units must operate normally for the successful system operation. The block diagram of a *k*-out-of-*n* unit network/system is shown in Figure 3.4 and each block in the diagram denotes a unit.

The parallel and series networks are special cases of this network for $k = 1$ and $k = n$, respectively.

With the aid of binomial distribution, for identical and independent units, we have the following expression for reliability of *k*-out-of-*n* unit network shown in Figure 3.4:

$$R_{k/n} = \sum_{j=k}^{n} \binom{n}{j} R^{j}(1-R)^{n-j} \tag{3.33}$$

where

$$\binom{n}{j} = \frac{n!}{(n-j)!\,j!} \tag{3.34}$$

$R_{k/n}$ is the *k*-out-of-*n* network/system reliability, and
R is the unit reliability.

For constant failure rates of the identical units, with the aid of Equations 3.11 and 3.33, we get

$$R_{k/n}(t) = \sum_{j=k}^{n} \binom{n}{j} e^{-j\lambda t}(1 - e^{\lambda t})^{n-j} \tag{3.35}$$

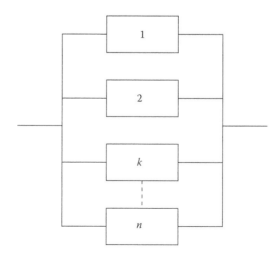

FIGURE 3.4
Block diagram of a *k*-out-of-*n* unit network.

where
$R_{k/n}(t)$ is the k-out-of-n network/system reliability at time t, and
λ is the unit constant failure rate.

By inserting Equation 3.35 into Equation 3.13, we obtain:

$$MTTF_{k/n} = \int_0^\infty \left[\sum_{j=k}^n \binom{n}{j} e^{-j\lambda t}(1 - e^{-\lambda t})^{n-j} \right] dt$$

$$= \frac{1}{\lambda} \sum_{j=k}^n \frac{1}{j} \tag{3.36}$$

where
$MTTF_{k/n}$ is the k-out-of-n network/system mean time to failure.

EXAMPLE 3.7

A robot system has four active, identical, and independent units in parallel. At least two units must operate normally for the successful operation of the robot system. Calculate the robot system's mean time to failure if the unit constant failure rate is 0.0008 failures per hour.
By inserting the given data values into Equation 3.36, we get

$$MTTF_{2/4} = \frac{1}{(0.0008)}\left[\frac{1}{2} + \frac{1}{3} + \frac{1}{4}\right]$$

$$= 1354.17 \text{ h}$$

Thus, the robot system mean time to failure is 1354.17 h.

3.4.4 Standby System

This is another reliability network/system/configuration in which only one unit works and m units are kept in their standby mode. The system contains a total of $(m + 1)$ units, and as soon as the operating unit fails, the switching mechanism detects the failure and turns on one of the standby units. The system fails when all its standby units fail.

Figure 3.5 shows the block diagram of a standby system with one operating and m standby units. Each block in the diagram denotes a unit. With the aid of Figure 3.5, for independent and identical units, perfect switching mechanism and standby units, and time-dependent unit failure rate, we have the following expression for reliability of the standby system [4,11]:

$$R_{ss}(t) = \frac{\sum_{j=0}^m \left[\left[\int_0^t \lambda(t)dt \right]^j e^{-\int_0^t \lambda(t)dt} \right]}{j!} \tag{3.37}$$

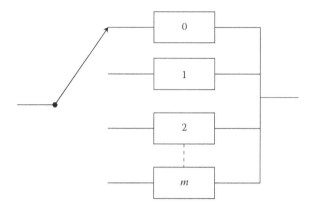

FIGURE 3.5
Block diagram of a standby system with one operating and m standby units.

where
     m is the number of standby units,
     $\lambda(t)$ is the unit time-dependent failure rate or hazard rate, and
     $R_{ss}(t)$ is the standby system reliability at time t.

For constant unit failure rate (i.e., $\lambda(t) = \lambda$), Equation 3.37 becomes

$$R_{ss}(t) = \frac{\sum_{j=0}^{m}(\lambda t)^{j} e^{-\lambda t}}{j!} \tag{3.38}$$

where
     λ is the unit constant failure rate.

By substituting Equation 3.38 into Equation 3.13, we get

$$MTTF_{ss} = \int_{0}^{\infty} \left[\frac{\sum_{j=0}^{m}(\lambda t)^{j} e^{-\lambda t}}{j!} \right] dt$$

$$= \frac{(m+1)}{\lambda} \tag{3.39}$$

where
     $MTTF_{ss}$ is the standby system's mean time to failure.

EXAMPLE 3.8

A robot system is composed of a standby system having two independent and identical units: one operating, the other on standby. The unit constant failure rate is 0.0004 failures per hour.

Calculate the standby system's reliability for a 100-h mission if the switching mechanism is perfect and the standby unit remains as good as new in its standby mode.

By substituting the specified data values into Equation 3.38, we obtain:

$$R_{ss}(100) = \sum_{j=0}^{1} \left[\frac{\{(0.0004)(100)\}^{j} e^{-(0.0004)(100)}}{j!} \right]$$
$$= 0.9992$$

Thus, the standby system reliability is 0.9992.

3.4.5 Bridge Network

Sometimes units or parts in robot systems may form a bridge network, as shown in Figure 3.6. Each block in Figure 3.6 denotes a unit, and all units are labeled with numerals.

For independent units, bridge-reliability for Figure 3.6 is given by [12]

$$\begin{aligned} R_b = {}& 2R_1R_2R_3R_4R_5 + R_1R_3R_5 + R_2R_3R_4 + R_2R_5 + R_1R_4 \\ & - R_1R_2R_3R_4 - R_1R_2R_3R_5 - R_2R_3R_4R_5 - R_1R_2R_4R_5 - R_3R_4R_5R_1 \end{aligned} \quad (3.40)$$

where
R_b is the bridge network's reliability, and
R_j is the unit j reliability, for $j = 1,2,3,4,5$.

For identical units, Equation 3.40 simplifies to

$$R_b = 2R^5 - 5R^4 + 2R^3 + 2R^2 \quad (3.41)$$

where
R is the unit reliability.

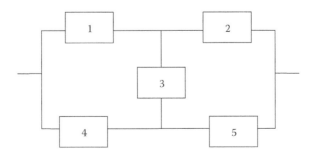

FIGURE 3.6
Five nonidentical unit bridge network.

For constant failure rates of all five units, with the aid of Equations 3.11 and 3.41, we obtain:

$$R_b(t) = 2e^{-5\lambda t} - 5e^{-4\lambda t} + 2e^{-3\lambda t} + 2e^{-2\lambda t} \tag{3.42}$$

where
λ is the unit constant failure rate.
$R_b(t)$ is the bridge network's reliability at time t.

By inserting Equation 3.42 into Equation 3.13, we obtain

$$MTTF_b = \int_0^\infty (2e^{-5\lambda t} - 5e^{-4\lambda t} + 2e^{-3\lambda t} + 2e^{-2\lambda t})dt$$

$$= \frac{49}{60\lambda} \tag{3.43}$$

where
$MTTF_b$ is the bridge network's mean time to failure.

EXAMPLE 3.9

Assume that in a robot system, five identical and independent units form a bridge network. Calculate the bridge network's reliability for a 250-h mission and mean time to failure, if each unit's constant failure rate is 0.0005 failures per hour.
By inserting the given data values into Equation 3.42, we obtain:

$$R_b(250) = 2e^{-5(0.0005)(250)} - 5e^{-4(0.0005)(250)} + 2e^{-3(0.0005)(250)} + 2e^{-2(0.0005)(250)}$$
$$= 0.9700$$

Similarly, substituting the specified data value into Equation 3.43 yields:

$$MTTF_b = \frac{49}{60(0.0005)}$$
$$= 1633.33 \text{ h}$$

Thus, the bridge network's reliability and mean time to failure are 0.9700 and 1633.33 h, respectively.

3.5 Need for Safety and the Role of Engineers with Respect to Safety

The desire to be safe and secure has always been an important concern to humans, going back to 2300 BC, with the laws of the Sumerian, Akkadian, Babylonian and Abyssinian empires (modern-day Iraq and Syria), including the codes of Urukagina, Ur-Nammu and Hammurabi, among others. The best known is the Code of Hammurabi [5,13].

Today, safety has become a very important issue because each year a vast number of people die and get seriously injured in workplace and other accidents around the globe. For example, in the United States alone in 1996, as per the United States National Safety Council, there were 93,400 deaths and a very large number of disabling injuries due to accidents [10,14]. The cost of these accidents was estimated to be about US$121 billion.

Factors that play a key role in demanding better safety include government regulations, an increasing number of law suits, and public pressure.

Today, engineering products or systems have become very complex and sophisticated. The matter of safety related to such products or systems has become a challenging issue for engineers who are pressured to complete new designs rapidly and at lower costs because of competition and other factors. Past experiences clearly indicate that this, in turn, generally leads to more design-related deficiencies, errors, and causes of accidents. Design-related deficiencies can contribute to or cause accidents.

As per [15], design-related deficiency may result because a designer/design:

- Fails to eliminate or reduce the occurrence of human error
- Offers confusing, wrong, or unfinished concepts/products
- Fails to prescribe an appropriate operational procedure in conditions where a hazard might exist
- Violates normal tendencies/capabilities of potential users
- Fails to properly warn of a potential hazard
- Incorporates weak warning mechanisms rather than providing a safe design for eradicating hazards
- Fails to provide an acceptable level of protection in a user's personal protective equipment
- Relies on users of product/item to avoid an accident
- Fails to foresee an unexpected application of an item/product/system or its potential consequences
- Creates an arrangement of operating controls and other devices that significantly increase reaction time during an emergency or is conducive to errors

- Creates an unsafe characteristic of an item/product/system
- Does not properly determine or consider the consequences of an action, error, failure, or omission
- Places an unreasonable level of stress on operators/users

3.6 Classifications of Product Hazards and Common Mechanical Injuries

There are many product-related hazards. As per [16], these may be grouped under the following six classifications:

- *Misuse-and-abuse hazards.* These hazards are concerned with product-usage by humans. Past experiences clearly indicate that product-misuse can lead to serious injuries. Product-abuse can also lead to hazardous situations or injuries; two examples of abuse are poor operating practices and lack of proper maintenance.
- *Electrical hazards.* Two principal components of electrical hazards are electrocution and shock. Past experiences indicate that the major electrical hazard to product/system/property stems from electrical faults, frequently referred to as short circuits.
- *Human-factor hazards.* These hazards are associated with poor design with regard to humans, more specifically, to their physical strength, height, intelligence, computational ability, education, weight, length of reach, visual angle, and so on.
- *Kinematic hazards.* These hazards are associated with situations where parts/items come together while moving and lead to pinching, crushing, or cutting of any object/item caught between them.
- *Environmental hazards.* These hazards may be categorized under two groups: internal hazards and external hazards. The internal hazards are associated with the changes in the surrounding environment that lead to internal damage in the product. This type of hazard can be eliminated or minimized by carefully considering, during the design phase, factors such as extremes of temperatures, vibrations, atmospheric contaminants, ambient noise level, and electromagnetic radiation.

  The external hazards are the hazards posed by the product/system during its life span. These hazards include service-life operational hazards, disposal hazards, and maintenance hazards.
- *Energy hazards.* These hazards may be categorized under two groups: kinetic energy hazards and potential energy hazards. The kinetic energy hazards pertain to items/parts that have energy because of their motion. Three examples of such items are fly wheels, loom

shuttles, and fan blades. Any object that interferes with their motion can experience extensive damage.

The potential energy hazards pertain to items/parts that store energy. Examples of such items/parts include springs, compressed-gas receivers, electronic capacitors, and counterbalancing weights. During the equipment-servicing processes, such hazards are important because stored energy can result in serious injury when released suddenly.

In the industrial sector, humans interact with various types of equipment to carry out tasks such as cutting, chipping, drilling, and stamping. There are various types of injuries that can occur in performing such tasks. Some of the commonly occurring injuries are as follows [5,10]:

- *Breaking injuries.* Generally, such injuries are associated with machines used to deform various types of engineering materials and can result in various types of fractures—simple fracture, oblique fracture, transverse fracture, and complete fracture.
- *Crushing injuries.* Such injuries occur when a body part is caught between two hard surfaces moving progressively together and crushing any object/item that comes between them.
- *Shearing injuries.* Such injuries pertain to shearing processes and include tragic damages such as amputation of fingers and hands.
- *Straining-and-spraining injuries.* Such injuries are generally associated with the use of machines or other tasks and include spraining of ligaments or straining of muscles.
- *Puncturing injuries.* Such injuries occur in a situation when an object penetrates sharply into an individual's body. In an industrial setting, generally, these injuries arise from punching machines because they have sharp tools.
- *Tearing and cutting injuries.* Such injuries occur when an individual's body part comes in contact with a sharp edge.

3.7 Organization Tasks for Product Safety and Safety Management Principles

An organization concerned with product safety performs various types of tasks. Some of these tasks are as follows [6,17]:

- Review safety-related customer complaints and field reports
- Review product-test reports to determine shortcomings or trends with regard to safety

- Review government and non-government requirements related to product safety
- Develop directives and programs related to product safety
- Review warning labels—regarding safety factors such as meeting requirements of, adequacy, and compatibility to warnings in the instruction manuals—that are to be placed on products
- Develop appropriate mechanisms by which the safety program can be monitored effectively
- Review mishaps and hazards in existing similar products/items for avoiding repetition of such hazards in new products/items
- Review the product to establish whether the potential hazards have been eradicated or controlled
- Determine if items such as emergency equipment, warning and monitoring devices, or protective equipment are required in handling or using the product
- Develop appropriate safety criteria based on applicable voluntary and governmental standards for use by the vendor, subcontractor, and company design professionals
- Provide appropriate assistance to designers in choosing alternative means to eradicate or control hazards or other safety-related problems in initial designs
- Review proposed product maintenance and operation documents with regard to safety
- Participate in reviewing accident-associated claims or recall actions by government agencies and recommend appropriate remedial actions for justifiable recalls or claims

Over the years, professionals working in the area of safety have developed various safety management principles. Some of these principles are as follows [6,18,19]:

- The function of safety is to find out and define the operational errors that result in accidents.
- The key to successful line-safety performance is management procedures that clearly factor in accountability.
- The causes that lead to unsafe behavior can be identified, controlled, and classified; the classifications of the causes include overload, the employee's decision to err, and traps.
- Safety should be managed just like any other function in an organization/company—more clearly, management should direct safety by setting goals that can be attained, and by planning, organizing, and controlling the attainment of such goals successfully.

- There are specific sets of circumstances that can predictably lead to severe injuries—nonproductive activities, certain construction conditions, high-energy sources, and abnormal nonroutine activities.

- The symptoms that indicate something is incorrect in the management system include accidents, unsafe actions, and unsafe conditions.

- Under most circumstances, normal human behavior is an unsafe behavior. Thus, it is the management team's responsibility as leaders to make appropriate changes to the environment that fosters unsafe behavior in order to encourage safe behavior.

- In developing a good safety system, the three main subsystems that must be considered with care are the behavioral, the managerial, and the physical subsystems.

- There is no single method to effectively achieve safety in an organization. But, for a safety system to be effective, it must meet certain criteria, such as: have the top management visibly showing its full support to safety, involve worker participation, be flexible, and so on.

3.8 Accident-Causation Theories

As per [5], there are many accident-causation theories. Two such theories are presented below.

3.8.1 The "Human Factors" Accident-Causation Theory

The "human factors" accident-causation theory is based on the assumption that accidents occur due to a chain of events directly or indirectly linked to human error. It consists of three main factors, shown in Figure 3.7, that lead to the occurrence of human error [5,20]. These three main factors are: inappropriate response/incompatibility; overload; and inappropriate activities.

The factor "inappropriate response/incompatibility" is an important factor for the occurrence of human error and some examples of inappropriate response by an individual are as follows [6,20]:

- A person completely disregards the specified safety procedures
- A person removes a safeguard from a machine for increasing output
- A person detects a hazardous condition, but takes no corrective action

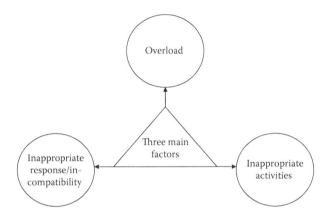

FIGURE 3.7
Three main factors that lead to the occurrence of human error.

The "overload" factor is concerned with an imbalance between the capacity of an individual at any time and the load he/she is carrying in a given state. The capacity of a person is the product of factors such as:

- The degree of training
- Fatigue
- Physical condition
- Stress
- State of mind
- Natural ability

The load carried by an individual is made up of tasks for which he/she has responsibility, along with additional burdens resulting from the situational factors (i.e., level of risk, unclear instructions, etc.), environmental factors (i.e., distractions, noise, etc.), and internal factors (i.e., personal problems, worry, emotional stress, etc.).

Finally, the factor of "inappropriate activities" is concerned with inappropriate activities undertaken due to human error, poor judgment about the degree of risk involved in a given task and subsequently acting on that misjudgment.

3.8.2 The "Domino" Accident-Causation Theory

This theory is encapsulated in 10 statements called the "Axioms of Industrial Safety" by H.W. Heinrick [21]. These 10 axioms are as follows [5,6,21]:

i. Most accidents are due to the unsafe acts of people.

ii. An unsafe act by a person or an unsafe condition does not always immediately lead to an accident/injury.

iii. Supervisors play a key role in the prevention of industrial accidents.

iv. The reasons why people commit unsafe acts can be useful in choosing appropriate corrective measures.

v. The severity of an injury is largely fortuitous, and the specific accident that caused it is generally preventable.

vi. There are two types of costs (i.e., direct and indirect) of an accident. Some examples of the direct costs are liability claims, hospital-related expenses, and compensation.

vii. The occurrence of injuries results from a completed sequence of numerous factors, the last one of which is the accident itself.

viii. Management should assume full responsibility for safety because it is in the best position for achieving the final results effectively.

ix. An accident can occur only when someone commits an unsafe act and/or there is a mechanical or physical hazard.

x. The most useful accident-prevention methods are analogous to the productivity and quality approaches.

According to Heinrich, the five specific factors in the sequence of events leading up to an accident are as follows [5,6]:

- *Ancestry and social environment.* In this factor, it is assumed that negative character traits such as recklessness, stubbornness, and avariciousness that might lead individuals to behave in an unsafe manner, can be acquired as a result of the social environment or surroundings or inherited through one's ancestry.

- *Fault of a person.* In this factor, it is assumed that negative character traits (whether inherited or acquired) such as ignorance of safety practices, excitability, recklessness, nervousness, and violent temper constitute proximate reasons for committing unsafe acts or for the existence of physical or mechanical hazards.

- *Unsafe act/physical or mechanical hazard.* In this factor, it is assumed that unsafe acts by people (i.e., starting equipment/machinery without warning, removing safeguards, standing under suspended loads) and physical or mechanical hazards (i.e., unguarded point of operation, unguarded gears, inadequate light, absence of guardrails) are the very direct causes for the occurrences of accidents.

- *Accident.* In this factor, it is assumed that events such as falls of humans and striking of humans by flying objects are the typical examples of accidents that cause injury.

- *Injury.* In this factor, it is assumed that the typical injuries that directly result from accidents include fractures and lacerations.

3.9 Problems

1. What are the three phases of a bathtub hazard rate curve? Discuss the causes of failures in each of these three phases.
2. Write down general equations for the following:
   i.   Reliability function
   ii. Hazard-rate function
3. Prove Equation 3.11 by using Equation 3.6.
4. Write down formulas to obtain mean time to failure by using the reliability function.
5. Assume that a robot system is composed of three independent and identical subsystems and the constant failure rate of each subsystem is 0.0005 failures per hour. All three subsystems must operate normally for the robot system to work successfully. Calculate the robot system reliability for a 50-hour mission, mean time to failure, and failure rate.
6. A robot system has four active, identical, and independent units in parallel. At least three units must operate normally for the successful operation of the robot system. Calculate the robot system's mean time to failure if the unit's constant failure rate is 0.006 failures per hour.
7. Prove Equation 3.31 by using Equation 3.35.
8. Discuss the "human factors" accident-causation theory.
9. List at least 10 tasks of a product-related safety organization.
10. Discuss product-hazard classifications.

References

1. Layman, W.J., Fundamental consideration in preparing a master plan, *Electric World*, Vol. 101, 1933, pp. 778–792.
2. Smith, S.A., Service reliability measured by probabilities of outage, *Electrical World*, Vol. 103, 1934, pp. 371–374.
3. Dhillon, B.S., *Power System Reliability, Safety, and Management*, Ann Arbor Science Publishers, Ann Arbor, Michigan, 1983.
4. Dhillon, B.S., *Design Reliability: Fundamentals and Applications*, CRC Press, Boca Raton, Florida, 1999.
5. Goetsch, D.L., *Occupational Safety and Health*, Prentice-Hall, Englewood Cliffs, New Jersey, 1996.

6. Dhillon, B.S., *Engineering Safety: Fundamentals, Techniques, and Applications*, World Scientific Publishing, River Edge, New Jersey, 2003.
7. Kapur, K.C., Reliability and maintainability, in *Handbook of Industrial Engineering*, Eds. G. Salvendy, John Wiley and Sons, New York, 1982, pp. 8.5.1–8.5.34.
8. Dhillon, B.S., Life distributions, *IEEE Transactions on Reliability*, Vol. 30, No. 5, 1981, pp. 457–460.
9. Shooman, M.L., *Probabilistic Reliability: An Engineering Approach*, McGraw-Hill Book Company, New York, 1968.
10. Dhillon, B.S., *Reliability, Quality, and Safety for Engineers*, CRC Press, Boca Raton, Florida, 2005.
11. Sandler, G.H., *System Reliability Engineering*, Prentice-Hall, Englewood Cliffs, New Jersey, 1963.
12. Lipp, J.P., Topology of switching elements versus reliability, *Transactions on IRE Reliability and Quality Control*, Vol. 7, 1957, pp. 21–34.
13. Ladon, J., Ed., *Introduction to Occupational Health and Safety*, National Safety Council (NSC), Chicago, Illinois, 1986.
14. National Safety Council, *Accidental Facts, Report*, Chicago, Illinois, 1996.
15. Hammer, W., Price, D., *Occupational Safety Management and Engineering*, Prentice-Hall, Upper Saddle River, New Jersey, 2001.
16. Hunter, T.A., *Engineering Design for Safety*, McGraw-Hill, New York, 1992.
17. Hammer, W., *Product Safety Management and Engineering*, Prentice-Hall, Englewood Cliffs, New Jersey, 1980.
18. Petersen, D., *Safety Management*, American Society of Safety Engineers, Des Plaines, Illinois, 1998.
19. Petersen, D., *Techniques of Safety Management*, McGraw-Hill, New York, 1971.
20. Heinrich, H.W., Petersen, D., Roos, N., *Industrial Accident Prevention*, McGraw-Hill, New York, 1980.
21. Heinrich, H.W., *Industrial Accident Prevention*, McGraw-Hill, New York, 1959.

4

Methods for Performing Reliability and Safety Analysis of Robot Systems

4.1 Introduction

Over the years, a large number of publications in the areas of reliability and safety have appeared in the form of articles in journals, related to conference proceedings, technical reports, and books [1–6]. Many of these publications report the development of various types of methods and approaches for analyses of reliability and safety. Some of these methods and approaches can be used quite effectively for analyzing both reliability and safety. The others are more confined to a specific area (i.e., reliability or safety).

Three examples of these methods and approaches that can be used to analyze both reliability and safety are failure modes and effect analysis (FMEA), the Markov method, and fault-tree analysis (FTA). FMEA was developed in the early 1950s to analyze engineering systems from the aspect of reliability. The Markov method is named after a Russian mathematician, Andrei A. Markov (1856–1922), and is a highly mathematical method that is frequently used for performing reliability and safety analyses in engineering systems. Finally, the FTA method was developed in the early 1960s to analyze rocket-launch control systems from the safety aspect.

Today, these three methods are used across many diverse areas to analyze various types of problems related, directly or indirectly, to reliability and safety [5–9]. This chapter presents a number of methods and approaches considered useful in analyzing the reliability or safety or both of robot systems. All have been extracted from the published literature in the areas of reliability and safety.

4.2 Failure Modes and Effect Analysis

FMEA is a widely used method for analyzing engineering systems for their reliability. The method may simply be described as an effective approach for analyzing each potential failure mode in the system to examine the effects of such failure modes on the entire system [10].

The history of FMEA may be traced back to early 1950s with the development of flight control systems, when the Bureau of Aeronautics of the U.S. Navy, in order to develop a procedure for controlling reliability in the detail design effort, developed a requirement known as Failure Analysis [11]. Subsequently, the term "Failure Analysis" was changed to FMEA. In the 1970s, the U.S. Department of Defense directed its effort to developing a military standard entitled "Procedures for Performing a Failure Mode, Effects, and Criticality Analysis (FMECA)" [12]. FMECA is an extended version of FMEA. More clearly, when FMEA is extended to group each potential failure effect with respect to its severity (this incorporates documenting catastrophic and critical failures), the method is called FMECA [5,6,13].

The seven main steps followed to perform FMEA are as follows [2,5]:

- *Step 1*: Define system boundaries and its associated detailed requirements.
- *Step 2*: List system subsystems and parts/components.
- *Step 3*: List each part's/component's failure modes, the description and the identification.
- *Step 4*: Assign the failure-occurrence probability/rate to each part's/component's failure mode.
- *Step 5*: List each failure mode's effect/effects on the subsystem, system, and plant.
- *Step 6*: Enter remarks for each failure mode.
- *Step 7*: Review each critical failure mode and take necessary action.

It is to be noted that there are many factors that must be explored with care prior to the implementation of FMEA. These factors include examination of each conceivable failure mode by the involved professionals, measuring FMEA cost/benefits, making decisions based on the risk priority number (RPN), and obtaining the engineer's approval and support [14].

Over the years, professionals involved with reliability-analysis have established certain guidelines/facts concerning FMEA. Five of these guidelines/facts are shown in Figure 4.1 [14]. Nonetheless, there are many benefits to performing FMEA. Some of the main ones are as follows [5,14]:

- Easy to understand
- A useful tool for improving communication among design interface personnel
- A systematic approach for classifying hardware failures
- Reduces development time and cost

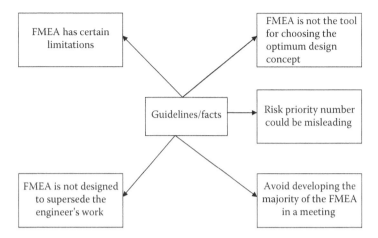

FIGURE 4.1
FMEA-related guidelines/facts.

- A useful approach that starts from the detailed level and works upward
- Identifies safety-related concerns for focus
- A visibility tool for management that also improves customer satisfaction
- A useful approach for comparing designs; reduces engineering changes
- A useful approach for safeguarding against repeating the same mistakes in the future
- A useful approach for more efficient test planning

Additional information on this method is available in [5,14].

4.3 The Markov Method

The Markov method is widely used to perform various types of reliability-analysis of engineering systems and is named after a Russian mathematician, Andrei A. Markov (1856–1922). The method is used frequently to model reparable systems with constant failure and repair rates.

The method is subject to the following assumptions [5,15]:

- All occurrences are independent of each other.

- The transitional probability from one system-state to another in the finite time interval Δt is given by $\alpha \Delta t$, where α is the transition rate (e.g., failure or repair rate) from one system state to another.
- The probability of more than one transition occurrence in the finite time interval Δt from one system state to another is negligible (e.g., $[\alpha \Delta t][\alpha \Delta t] \rightarrow 0$).

The following example demonstrates the application of this method.

EXAMPLE 4.1

Assume that a robot system can either be in an operating or a failed state and its constant failure and repair rates are α_r and μ_r, respectively. The robot-system state-space diagram is shown in Figure 4.2. The numerals in the box and circle denote the robot-system states. Develop expressions for the robot system, with the aid of the Markov method, time-dependent and steady-state availabilities and unavailabilities, reliability, and mean time to failure.

With the aid of the Markov method, we write down the following equations for states 0 and 1, respectively shown in Figure 4.2:

$$P_0(t + \Delta t) = P_0(t)(1 - \alpha_r \Delta t) + P_1(t)\mu_r \Delta t \qquad (4.1)$$

$$P_1(t + \Delta t) = P_1(t)(1 - \mu_r \Delta t) + P_0(t)\alpha_r \Delta t \qquad (4.2)$$

where
  t is time,
  $\alpha_r \Delta t$ is the probability of robot system failure in finite time interval Δt,
  $(1 - \alpha_r \Delta t)$ is the probability of no robot system failure in finite time interval Δt,
  $\mu_r \Delta t$ is the probability of robot system repair in finite time interval Δt,
  $(1 - \mu_r \Delta t)$ is the probability of no robot system repair in finite time interval Δt,
  $P_j(t)$ is the probability that the robot system is in state j at time t, for $j = 0,1$,
  $P_0(t + \Delta t)$ is the probability of the robot system being in operating state 0 at time $(t + \Delta t)$, and

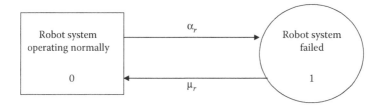

FIGURE 4.2
Robot system state-space diagram.

$P_1(t + \Delta t)$ is the probability of the robot system being in failed state 1 at time $(t + \Delta t)$.

From Equation 4.1, we write

$$P_0(t + \Delta t) = P_0(t) - P_0(t)\alpha_r \Delta t + P_1(t)\mu_r \Delta t \qquad (4.3)$$

From Equation 4.3, we get

$$\lim_{\Delta t \to 0} \frac{P_0(t + \Delta t) - P_0(t)}{\Delta t} = -P_0(t)\alpha_r + P_1(t)\mu_r \qquad (4.4)$$

Thus, from Equation 4.4, we obtain

$$\frac{dP_0(t)}{dt} + P_0(t)\alpha_r = P_1(t)\mu_r \qquad (4.5)$$

Similarly, using Equation 4.2, we obtain

$$\frac{dP_1(t)}{dt} + P_1(t)\mu_r = P_0(t)\alpha_r \qquad (4.6)$$

At time $t = 0$, $P_0(0) = 1$, and $P_1(0) = 0$.
By solving Equations 4.5 and 4.6, we get [5]

$$P_0(t) = \frac{\mu_r}{(\alpha_r + \mu_r)} + \frac{\alpha_r}{(\alpha_r + \mu_r)} e^{-(\alpha_r + \mu_r)t} \qquad (4.7)$$

$$P_1(t) = \frac{\alpha_r}{(\alpha_r + \mu_r)} - \frac{\alpha_r}{(\alpha_r + \mu_r)} e^{-(\alpha_r + \mu_r)t} \qquad (4.8)$$

Thus, the robot system time-dependent availability and unavailability, respectively, are

$$A_r(t) = P_0(t) = \frac{\mu_r}{(\alpha_r + \mu_r)} + \frac{\alpha_r}{(\alpha_r + \mu_r)} e^{-(\alpha_r + \mu_r)t} \qquad (4.9)$$

and

$$UA_r(t) = P_1(t) = \frac{\alpha_r}{(\alpha_r + \mu_r)} - \frac{\alpha_r}{(\alpha_r + \mu_r)} e^{-(\alpha_r + \mu_r)t} \qquad (4.10)$$

where
$A_r(t)$ is the robot system availability at time t and
$UA_r(t)$ is the robot system unavailability at time t.

By letting time t go to infinity in Equations 4.9 and 4.10, we obtain [5]:

$$A_r = \lim_{t \to \infty} A_r(t) = \frac{\mu_r}{\alpha_r + \mu_r} \qquad (4.11)$$

and

$$UA_r = \lim_{t \to \infty} UA_r(t) = \frac{\alpha_r}{\alpha_r + \mu_r} \qquad (4.12)$$

where
A_r is the robot system steady-state availability and
UA_r is the robot system steady-state unavailability.

For $\mu_r = 0$, from Equation 4.9 we obtain

$$R_r(t) = P_0(t) = e^{-\alpha_r t} \qquad (4.13)$$

where
$R_r(t)$ is the robot system reliability at time t.

By integrating Equation 4.13 over the time interval $[0,\infty]$, we obtain the following equation for the robot-system mean time to failure [5]:

$$MTTF_r = \int_0^{\infty} e^{-\alpha_r t} dt$$

$$= \frac{1}{\alpha_r} \qquad (4.14)$$

where
$MTTF_r$ is the robot system mean time to failure.

Thus, the robot-system time-dependent and steady-state availabilities and unavailabilities, reliability, and mean time to failure are given by Equations 4.9 through 4.14.

EXAMPLE 4.2

Assume that a robot system's constant failure and repair rates are 0.0006 failures per hour, and 0.005 repairs per hour, respectively. Calculate the robot system steady-state unavailability and unavailability during a 20 h mission.

By inserting the specified data values into Equations 4.12 and 4.10, we get

$$UA_r = \frac{0.0006}{0.0006 + 0.005} = 0.1071$$

and

$$UA(20) = \frac{0.0006}{(0.0006 + 0.005)} - \frac{0.0006}{(0.0006 + 0.005)} e^{-(0.0006+0.005)(20)}$$
$$= 0.013$$

Thus, the robot-system steady-state unavailability and unavailability during a 20 h mission are 0.1071 and 0.113, respectively.

4.4 Fault Tree Analysis

This method was developed in the early 1960s at the Bell Telephone Laboratories for performing reliability-analysis of the Minuteman Launch Control System [2,5]. Today, it is widely used to evaluate the reliability of engineering systems in their design and development, particularly in the area of nuclear power generation.

A fault tree may simply be described as a logical representation of the relationship of primary fault events that lead to the occurrence of a stated undesirable event called the "top event" and is depicted using a tree structure with logic gates such as OR and AND. Some of the purposes of performing FTA are as follows [5]:

- Understanding the level of protection that the design concept provides against failures
- Understanding the functional relationship of system failures
- Meeting jurisdictional requirements
- Identifying critical areas and potential for cost-effective improvements

FTA begins by identifying an undesirable event, called the "top event," associated with a system under consideration. Fault events that can lead to the occurrence of the top event are generated and connected by logic operators such as OR and AND. The OR gate provides a true output (i.e., fault) if one or more inputs (i.e., faults) are true, whereas the AND gate provides a true output (i.e., fault) if all the inputs (i.e., faults) are true.

The construction of a fault tree proceeds by generating fault events in a successive manner until the fault events need not be developed any further. These fault events are known as primary or basic events. During the process of fault tree construction, one successively asks the question: "How could

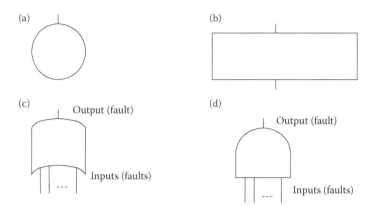

FIGURE 4.3
Fault tree symbols: (a) circle, (b) rectangle, (c) OR gate, (d) AND gate.

this fault event occur?" Basic symbols used in constructing fault trees are shown in Figure 4.3 [2,5].

All the four symbols shown in Figure 4.3 are described below (it is to be noted that for the sake of clarity OR and AND gates are described again).

- *Circle*: It denotes a basic or primary fault event (e.g., failure of an elementary part or component). The parameters of the event are probability of occurrence, failure rate, and repair rate; whose values are normally obtained from empirical data.
- *Rectangle*: It denotes a fault event that occurs from the logical combination of fault events through the input of a logic gate such as AND and OR.
- *OR gate*: It represents an output fault event that occurs if one or more of the input fault events occur.
- *AND gate*: It represents an output event that occurs only if all of the input fault events occur.

Information on other symbols used in conducting FTA is available in [2,5].

EXAMPLE 4.3

Assume that a windowless room containing a robot system has three light bulbs (i.e., *X,Y,Z*) and one switch. The switch can only fail to close. Develop a fault tree for the undesirable event (i.e., top fault event) "dark robot system room" (i.e., no light in the room containing the robot system) using symbols presented in Figure 4.3.

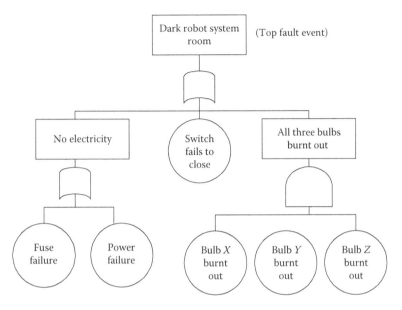

FIGURE 4.4
A fault tree for the top fault event: dark robot system room.

In this case, the robot system room can only be dark if the switch fails to close, if all the three light bulbs burn out, or if there is no incoming electricity. A fault tree for the example is shown in Figure 4.4.

4.4.1 Probability Evaluation of Fault Trees

Under certain circumstances, it may be necessary to predict the occurrence probability of a certain event (e.g., no light in the room containing the robot system). Before this could be achieved by using the FTA approach, the calculation of the probability of occurrence of the output-fault events of logic gates is necessary.

Thus, the probability of occurrence of the output-fault event of an AND gate is expressed by [2,5]

$$P_a(Y) = \prod_{j=1}^{m} P(Y_j) \qquad (4.15)$$

where
$P_a(Y)$ is the probability of occurrence of AND gate's output-fault event Y,
m is the number of independent input-fault events and
$P(Y_j)$ is the probability of occurrence of input-fault event Y_j, for $j = 1,2,3,\ldots,m$.

Similarly, the probability of occurrence of the output-fault event of an OR gate is given by [2,5]

$$P_o(Y) = 1 - \prod_{j=1}^{m}(1 - P(Y_j))$$ (4.16)

where

$P_0(Y)$ is the probability of occurrence of OR gate's output-fault event Y.

EXAMPLE 4.4

Assume that in Figure 4.4 the probabilities of occurrence of fault events— power failure, fuse failure, switch fails to close, burnt-out bulb X, burnt-out bulb Y, and burnt-out bulb Z—are 0.02, 0.03, 0.04, 0.05, 0.05, and 0.05, respectively. With the aid of Equations 4.15 and 4.16, calculate the probability of occurrence of the top fault event "dark robot system room" and redraw Figure 4.4 with the given and calculated probability values.

By substituting the given data values into Equation 4.15, the probability of occurrence of the event "all three bulbs burnt out" is

$$P_b = (0.05)(0.05)(0.05)$$
$$= 0.000125$$

where

P_b is the probability of occurrence of the event "all three bulbs burnt out."

Similarly, by inserting the specified data values into Equation 4.16, the probability of occurrence of the event "no electricity" is

$$P_n = (0.03) + (0.02) - (0.03)(0.02)$$
$$= 0.0494$$

where

P_n is the probability of occurrence of the event "no electricity."

By inserting the above calculated values and the specified data value into Equation 4.16, the probability of occurrence of the top fault event "dark robot system room" is

$$P_d = 1 - (1 - 0.0494)(1 - 0.04)(1 - 0.000125)$$
$$= 0.0875$$

Thus, the probability of occurrence of the top fault event "dark robot system room" is 0.0875. Figure 4.4 with the given data and calculated values is shown in Figure 4.5.

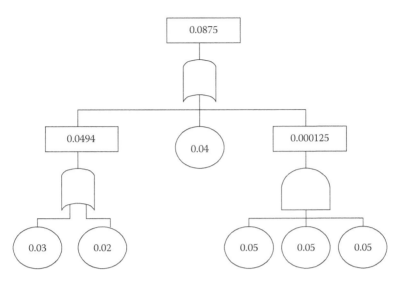

FIGURE 4.5
A fault tree with the given and calculated event occurrence probability values.

4.4.2 FTA Advantages and Disadvantages

There are many advantages and disadvantages of performing FTA. Some of its advantages are as follows [2,5]:

- Allows concentration on one specific failure at a time
- Offers a graphic aid for management
- Useful to handle complex systems more easily
- Requires the analyst to understand thoroughly the system under consideration before starting the analysis
- Useful to provide options for management and others to perform either qualitative or quantitative analysis
- Useful to highlight failures deductively and to provide insight into the system behavior

In contrast, some of the disadvantages of the FTA are as follows [2,5]:

- A time-consuming and costly method
- The end results are difficult to check
- It considers parts or components in either working or failed state (i.e., the partial failure states of the parts or components are difficult to handle)

Additional information on this method is available in [2,5].

4.5 Technique of Operations Review

Technique of operations review (TOR) may simply be described as a hands-on analytical approach to identify the root system causes of an operation failure; it was developed in the early 1970s by D.A. Weaver of the American Society of Safety Engineers [6,16]. The method seeks to highlight systemic causes for an adverse incident rather than assigning blame with respect to safety. Thus, it permits management personnel and workers to work jointly to analyze workplace-related accidents, failures, and incidents.

The method is activated by an adverse incident occurring at a certain point in time and location involving certain individuals, and makes use of a worksheet containing simple terms that require yes/no responses. It is to be noted that the method is not a hypothetical process and it demands a systematic evaluation of the circumstances surrounding the incident in question. Ultimately, TOR highlights how the company/organization could have prevented the occurrence of the accident.

The following steps are associated with the method [6,17]:

- *Step 1*: Form the TOR team with appropriate members belonging to all concerned areas.
- *Step 2*: Hold a roundtable session to impart common knowledge to all members of the TOR team.
- *Step 3*: Identify one important systemic factor that played a key role in the occurrence of the accident/incident. It is to be noted that this factor must be based on the consensus of all members of the TOR team and it serves as a starting point for further investigation to follow.
- *Step 4*: Make use of the team consensus in responding to a sequence of yes/no options.
- *Step 5*: Evaluate the identified factors, ensuring that there is clear-cut consensus among the team personnel with respect to each factor.
- *Step 6*: Prioritize the contributory factors by starting with the one considered the most serious.
- *Step 7*: Develop necessary preventive/corrective strategies with respect to each and every contributory factor.
- *Step 8*: Implement the strategies.

Finally, it is to be noted that the main strength of this method is the involvement of line personnel in the analysis and its main weakness is that it is an after-the-fact process.

4.6 Hazard and Operability Analysis

This method is considered a powerful tool to identify safety-related problems prior to the availability of complete data concerning an item and it was developed for application in the chemical industry sector [18,19]. Three basic objectives of Hazard and Operability (HAZOP) analysis are as follows [17,20,21]:

- To review each part of the facility/process for determining how deviations from the intended design can happen
- To decide whether the deviations can lead to operating problems/hazards
- To produce a complete description of the facility/process

The seven steps followed to perform HAZOP analysis are shown in Figure 4.6 [17,19,22].

Finally, it is added that this method has basically the same weaknesses as FMEA discussed earlier in the chapter. For example, both HAZOP analysis and FMEA predict problems that are connected to process/system faults/failures, but overlook the factoring in of human error into the equation. This is an important weakness because, frequently, human error is a factor in accident occurrences.

4.7 Interface Safety Analysis

This method is concerned with determining the incompatibilities between subsystems and assemblies of an equipment/product that could lead to accidents. The method establishes that distinct units/parts can be integrated into a viable system and that normal functioning of an individual unit/part will not impair the performance of or damage another unit/part or the complete equipment/system. Although the method considers various relationships, they can be grouped under the following three classifications [6,23]:

- *Classification I: Functional relationships.* These relationships are concerned with multiple items or units. For example, in situations where outputs of an item or unit constitute the inputs to the downstream item(s) or unit(s), any error in outputs and inputs may result in damage to the downstream item(s) or unit(s), thereby creating a safety problem or hazard. Such outputs could be in conditions such as erratic outputs, degraded outputs, excessive outputs, unprogrammed outputs, and zero outputs.

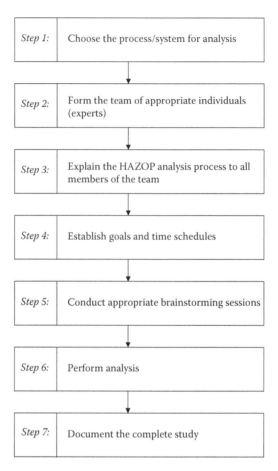

FIGURE 4.6
HAZOP analysis steps.

- *Classification II: Flow relationships.* These relationships are concerned with two or more items or units. For example, the flow between two items may involve electrical energy, steam, water, air, lubricating oil, or fuel. In addition, the flow could also be unconfined, such as heat radiation from one item to another.

  Often, the problems experienced with many products or items include the proper flow of fluids and energy from one unit to another unit through confined passages, consequently leading to safety problems. The causes of flow problems include total or partial interconnection failure and faulty connections between units. In the case of fluids, from the safety perspective, the factors that must be considered with utmost care include contamination, flammability, lubricity, loss of pressure, odor, and toxicity.

- *Classification III: Physical relationships.* These are concerned with the physical aspects of products/items. For example, two products or items might be well designed and manufactured and operate quite well individually, but they may have problems in fitting together because of dimension differences, or there may be other incompatibilities that may result in safety problems. Some examples of the other problems are as follows:
  - A small clearance between units; thus, during the removal process, the units may be damaged
  - Situation where it is impossible to tighten, join, or mate parts/units properly
  - Situation where access to or egress from system/equipment is restricted or impossible

4.8 Probability Tree Method

The probability tree method is used for performing task analysis by diagrammatically representing critical human actions and other events concerning the system under consideration. Often, the method is used to perform task analysis in the technique for human error rate prediction (THERP) [2,8]. In this method, diagrammatic task analysis is denoted by the probability tree branches. More clearly, the tree's branching limbs denote the outcome (i.e., success or failure) of each event and each branch is assigned a value for the probability of occurrence.

Some of the benefits of the probability-tree method are as follows [2,24]:

- Decreases the probability of errors in computation because of simplification in the computational process
- Includes, with some modifications, factors such as interaction stress, interaction effects, and emotional stress
- A useful visibility tool

The application of this method is demonstrated through the following two examples:

EXAMPLE 4.5

Assume that a robot system maintenance person performs two independent safety-related tasks: x and y. Each of these two tasks can be performed either correctly or incorrectly and task x is carried out before task y.

Develop a probability tree and obtain an expression for the probability of not successfully accomplishing the overall mission by the robot system maintenance person.

In this case, the robot system maintenance person performs task x correctly or incorrectly and then proceeds to perform task y. Task y can also be performed correctly or incorrectly by the robot system maintenance person. Figure 4.7 shows this entire scenario.

The symbols used in Figure 4.7 are defined below.

x is the task x performed correctly.
$\bar{x}$ is the task x performed incorrectly.
y is the task y performed correctly.
$\bar{y}$ is the task y performed incorrectly.

By examining Figure 4.7, it is concluded that there is a total of three possibilities (i.e., $x\bar{y}$, $\bar{x}y$, and $\bar{x}\bar{y}$) for not successfully accomplishing the overall mission by the robot system maintenance person. Thus, the probability of not successfully accomplishing the overall mission by the robot system maintenance person is given by

$$
\begin{aligned}
P_f &= P(x\bar{y}) + P(\bar{x}y) + P(\bar{x}\bar{y}) \\
&= P_x P_{\bar{y}} + P_{\bar{x}} P_y + P_{\bar{x}} P_{\bar{y}}
\end{aligned}
\tag{4.17}
$$

where
P_f is the probability of not successfully accomplishing the overall mission by the robot system maintenance person.
P_x is the probability of performing the independent task x correctly by the robot system maintenance person.
P_y is the probability of performing the independent task y correctly by the robot system maintenance person.
$P_{\bar{x}}$ is the probability of performing the independent task x incorrectly by the robot system maintenance person.

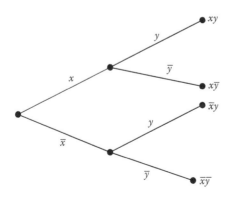

FIGURE 4.7
Probability tree diagram for the robot system maintenance person performing tasks x and y.

$P_{\bar{y}}$ is the probability of performing the independent task y incorrectly by the robot system maintenance person.

Thus, Equation 4.17 is the expression for the probability of not successfully accomplishing the overall mission by the robot system maintenance person.

EXAMPLE 4.6

With the aid of Figure 4.7, for Example 4.5, develop a probability expression for successfully accomplishing the overall mission by the robot system maintenance person. In addition, calculate the probability of not successfully accomplishing the overall mission by the robot system maintenance person, if the probabilities of performing tasks x and y correctly are 0.85 and 0.95, respectively.

Thus, with the aid of Figure 4.7, the probability of successfully accomplishing the overall mission by the robot system maintenance person is

$$P_s = P(xy)$$
$$= P_x P_y \tag{4.18}$$

where
    P_s is the probability of successfully accomplishing the overall mission by the robot system maintenance person.

Because the $P_{\bar{x}} = 1 - P_x$ and $P_{\bar{y}} = 1 - P_y$, Equation 4.17 reduces to

$$P_f = 1 - P_x P_y \tag{4.19}$$

By inserting the specified data values into Equation 4.19, we get

$$P_f = 1 - (0.85)(0.95)$$
$$= 0.1925$$

Thus, the probability of not successfully accomplishing the overall mission by the robot system maintenance person is 0.1925.

4.9 Problems

1. What are the main advantages of FMEA?
2. What is the difference between FMECA and FMEA? Describe FMEA.
3. Prove Equations 4.7 and 4.8 by using Equations 4.5 and 4.6.
4. Assume that a robot system's constant failure and repair rates are 0.0004 failures per hour and 0.002 repairs per hour, respectively. Calculate the robot system steady-state unavailability and availability during a 15 h mission.

5. What are the purposes of performing FTA?

6. Describe the following four items used in performing FTA:

   a. Circle

   b. OR gate

   c. Rectangle

   d. AND gate

7. Describe TOR.

8. Compare HAZOP analysis with FMEA.

9. Assume that a robot system maintenance person performs three independent safety-related tasks: x, y, and z. Each of these three tasks can be performed either correctly or incorrectly and task x is performed before task y, and task y before task z.

   Develop a probability-tree and obtain an expression for the probability of not successfully accomplishing the overall mission by the robot system maintenance person.

10. Describe ISA.

References

1. Countinho, W.E., *Failure Effect Analysis*, Trans. New York Academy of Sciences, Series II, 1963–1964, pp. 564–584.

2. Dhillon, B.S., Singh, C., *Engineering Reliability: New Techniques and Applications*, John Wiley and Sons, New York, 1981.

3. United States Nuclear Regulatory Commission, *Fault Tree Handbook, Report No. NUREG-0492*, Washington, DC, 1981.

4. Hammer, W., Price, D., *Occupational Safety Management and Engineering*, Prentice-Hall, Upper Saddle River, New Jersey, 2001.

5. Dhillon, B.S., *Design Reliability: Fundamentals and Applications*, CRC Press, Boca Raton, Florida, 1999.

6. Dhillon, B.S., *Engineering Safety: Fundamentals, Techniques, and Applications*, World Scientific Publishing, River Edge, New Jersey, 2003.

7. Dhillon, B.S., *Patient Safety: An Engineering Approach*, CRC Press, Boca Raton, Florida, 2012.

8. Dhillon, B.S., *Human Reliability: With Human Factors*, Pergamon Press, New York, 1986.

9. Dhillon, B.S., *Human Reliability, Error, and Human Factors in Engineering Maintenance: With Reference to Aviation and Power Generation*, CRC Press, Boca Raton, Florida, 2009.

10. Omdahl, T.P., Editor, *Reliability, Availability, and Maintainability (RAM) Dictionary*, American Society for Quality Control (ASQC) Press, Milwaukee, Wisconsin, 1988.

11. Bureau of Naval Weapons, U.S. Department of the Navy, *General Specification for Design, Installation, and Test of Aircraft Flight Control Systems*, MIL-F-18372 (Aer), Para. 3.5.2.3., Washington, DC.
12. U.S. Department of Defense, *Procedures for Performing a Failure Mode, Effects, and Criticality Analysis*, MIL-STD-1629, Washington, DC, 1980.
13. Jordan, W.E., Failure modes, effects, and criticality analyses, *Proceedings of the Annual Reliability and Maintainability Symposium*, 1972, pp. 30–37.
14. Palady, P., *Failure Modes and Effects Analysis*, PT Publications, West Palm Beach, Florida, 1995.
15. Shooman, M.L., *Probabilistic Reliability: An Engineering Approach*, McGraw Hill Book Company, New York, 1968.
16. Hallock, R.G., Technique of operations review analysis: Determines causes of accident/incident, *Safety and Health*, Vol. 60, No. 8, 1991, pp. 38–39.
17. Goetsch, D.L., *Occupational Safety and Health*, Prentice-Hall, Englewood Cliffs, New Jersey, 1996.
18. *Guidelines for Hazard Evaluation Procedures*, American Institute of Chemical Engineers, New York, 1985.
19. Dhillon, B.S., *Reliability, Quality, and Safety for Engineers*, CRC Press, Boca Raton, Florida, 2005.
20. Roland, H.E., Moriarty, B., *System Safety Engineering and Management*, John Wiley and Sons, New York, 1983.
21. Gloss, D.S., Wardle, M.G., *Introduction to Safety Engineering*, John Wiley and Sons, New York, 1984.
22. Canadian Standards Association, *Risk Analysis Requirements and Guidelines, Report No. CAN/CSA-Q 634-91*, 1991; available from the Canadian Standards Association, 178 Rexdale Blvd., Rexdale, Ontario, Canada.
23. Hammer, W., *Product Safety Management and Engineering*, Prentice-Hall, Englewood Cliffs, New Jersey, 1980.
24. Swain, A.D., An error-cause removal program for industry, *Human Factors*, Vol. 12, 1973, pp. 207–221.

5

Robot Reliability

5.1 Introduction

The topic of robot reliability is quite complicated as there are many interlocking variables in evaluating and accomplishing numerous reliability levels. A robot installation is only considered successful if it is safe and reliable. A robot with poor reliability results in problems such as high maintenance-related cost, inconvenience, and unsafe conditions.

Needless to say, the *American National Standard for Industrial Robots and Robot Systems—Safety Requirements* developed by the American National Standards Institute (ANSI) clearly states that the design and construction of robots has to be in such a way that any single, reasonably foreseeable malfunction will not result in any hazardous motion of the robot. As many different types of parts (e.g., electrical, electronic, mechanical, hydraulic, and pneumatic) are used in robots, this makes the task of producing highly reliable robots very challenging. Furthermore, the environments in which the robots have to function may be quite harsh and can vary quite significantly from one application to another even for identical models.

This chapter presents various important aspects of robot reliability.

5.2 Classifications of Robot Failures and Their Causes and Corrective Measures

Robot failures can be categorized under the following four classifications [1–3]:

- *Classification I: Random component failures.* These are failures that occur unpredictably during the useful life of components. Some of the reasons for the occurrence of such failures are low safety factors, undetectable defects, unavoidable failures, and unexplainable causes. Some of the methods presented in Chapter 4 and in [4] can be used to reduce the occurrence of such failures.

- *Classification II: Human errors.* These are due to personnel who design, manufacture, test, operate, and maintain robots. Some of the causes for the occurrence of human errors are as follows:

  - Poor training of operating and maintenance personnel
  - Poor equipment design
  - Improper tools
  - Task complexities
  - Poorly written operating and maintenance procedures
  - Inadequate lighting in the work area
  - High temperature in the work area

  Thus, human errors may be divided into categories such as design errors, operating errors, maintenance errors, inspection errors, installation errors, and assembly errors. Some of the methods that can be used to reduce the occurrence of human errors are fault tree analysis, error cause removal program, man–machine systems analysis, and quality control circles. The first method (i.e., fault-tree analysis) is described in Chapter 4 and the remaining three methods are described in [5].

- *Classification III: Systematic hardware faults.* These are failures that occur due to unrevealed mechanisms present in the root system design. Some of the reasons for the occurrence of such faults are peculiar wrist orientations, unusual joint-to-straight-line mode transition, and failure to make the necessary environment-related provisions in the initial design.

  Some of the methods that can be used to reduce the occurrence of robot-related systematic hardware failures are the inclusion of sensors in the system to detect the loss of pneumatic pressure, line voltage, or hydraulic pressure; and the employment of sensors to detect excessiveness of force, speed, servo errors, acceleration, and temperature.

  Several methods useful to reduce systematic hardware failures are described in [4,6].

- *Classification IV: Software failures/errors.* These are associated with software concerned with robots. In robots, software faults/errors/ failures can happen in the embedded software or the controlling software and application software. As per one study reported in [7], over 60% of software errors are made during the requirement and design phase of the process as opposed to less than 40% during the coding phase or process.

Redundancy, even though it is expensive, is probably the best solution to protect against software errors or failures. Also, the application of approaches such as failure mode and effect analysis, testing, and fault tree analysis can be helpful to reduce software failures/errors. Furthermore, there are many software reliability models that can be used to evaluate reliability when the software in question is put into operational use [4,6–8].

5.3 Robot Effectiveness Dictating Factors and Robot Reliability Survey Results

There are many factors dictating effectiveness of robots. Some of these factors are as follows [3,9]:

- The robot's mean time between failures
- The robot's mean time to repair
- The percentage of time the robot is available for operations
- The percentage of time the robot operates normally
- Quality and availability of the robot repair equipment and facilities
- Rate of the availability of the required spare parts or components
- Quality and availability of personnel required for keeping the robot in operating state
- The relative performance of the robot under extreme conditions

Reference [10] reported the results of a study of robot reliability based on surveys of 37 robots of four different designs used in three different companies A, B, and C; covering 21, 932 robot production hours. These companies (i.e., A, B, and C) reported 47, 306, 155 cases of reliability-related problems of robots, respectively, of which the corresponding 27, 35, and 1 cases did not contribute to downtime. More clearly, robot downtime as a proportion of production time for these companies (i.e., A, B, and C) was 1.8%, 13.6%, and 5.1%, respectively.

Approximate mean time to robot failure (*MTTRF*) and mean time to robot-related problems (*MTTRP*) in hours for companies A, B, and C are presented in Table 5.1.

It is to be noted that among these three companies, there is a wide variation of *MTTRF* and *MTTRP*. More clearly, the lowest and highest *MTTRF* and *MTTRP* levels are 40, 2596, 15 and 221 h, respectively.

TABLE 5.1

Approximate Figures for Mean Time to Robot Failure (*MTTRF*)
and Mean Time to Robot-Related Problems (*MTTRP*)

No.	Company	*MTTRF* (h)	*MTTRP* (h)
1	A	2596	221
2	B	284	30
3	C	40	15

5.4 Robot-Related Reliability Measures

There are various types of robot-related reliability measures. Four of these
are presented below [3,4,11].

5.4.1 Mean Time to Robot-Related Problems

This is the average productive robot time before the occurrence of a robot-
related problem and is expressed by

$$MTTRP = \frac{RPH - DDTRP}{NRP} \tag{5.1}$$

where
  $MTTRP$ is the mean time to robot-related problems.
  RPH is the robot production hours.
  NRP is the number of robot-related problems.
  $DDTRP$ is the downtime due to robot-related problems expressed in hours.

EXAMPLE 5.1

Assume that the annual robot production hours and downtime due
to robot-related problems, at an industrial facility, are 6000 and 500 h,
respectively. During the 1-year period, there were 25 robot-related problems. Calculate the average time to robot-related problems.

By substituting the given data values into Equation 5.1, we get

$$MTTRP = \frac{6000 - 500}{25}$$
$$= 220 \text{ h}$$

Thus, the average time to robot-related problems is 220 h.

5.4.2 Mean Time to Robot-Related Failure

Mean time to robot-related failure can be obtained by using either of the following three expressions:

$$MTTF_r = \frac{RPH - DDTRF}{NRF} \tag{5.2}$$

$$MTTF_r = \int_0^\infty R_r(t)dt \tag{5.3}$$

$$MTTF_r = \lim_{s \to 0} R_r(s) \tag{5.4}$$

where
$MTTF_r$ is the robot's mean time to failure.
$DDTRF$ is the downtime due to robot-related failure expressed in hours.
NRF is the number of robot failures.
$R_r(t)$ is the robot reliability at time t.
s is the Laplace transform variable.
$R_r(s)$ is the Laplace transform of the robot reliability function, $R_r(t)$.

EXAMPLE 5.2

Assume that the annual production hours of a robot and its annual downtime due to failures are 4000 and 200 h, respectively. During that period, the robot failed five times. Calculate the mean time to robot-related failure.

By inserting the specified data values into Equation 5.2, we obtain

$$MTTF_r = \frac{4000 - 200}{5}$$
$$= 760 \text{ h}$$

Thus, the mean time to robot-related failure is 760 h.

EXAMPLE 5.3

Assume that the constant failure rate, λ_r, of a robot is 0.0002 failures per hour and its reliability is expressed by

$$R_r(t) = e^{-\lambda_r t}$$
$$= e^{-(0.0002)t} \tag{5.5}$$

where
$R_r(t)$ is the robot reliability at time t.

Calculate the mean time to robot-related failure with the aid of Equations 5.3 and 5.4 and comment on the final result.

By inserting Equation 5.5 into Equation 5.3, we get

$$
\begin{aligned}
\text{MTTF}_r &= \int_0^\infty e^{-(0.0002)t}\,dt \\
&= \frac{1}{0.0002} \\
&= 5000\ \text{h}
\end{aligned}
$$

By taking the Laplace transform of Equation 5.5, we obtain

$$
R_r(s) = \frac{1}{(s + 0.0002)} \tag{5.6}
$$

By substituting Equation 5.6 into Equation 5.4, we get

$$
\begin{aligned}
\text{MTTF}_r &= \lim_{s \to 0} \frac{1}{(s + 0.0002)} \\
&= \frac{1}{0.0002} \\
&= 5000\ \text{h}
\end{aligned}
$$

In both cases, the final result (i.e., $\text{MTTF}_r = 5000$ h) is the same. It proves that Equations 5.3 and 5.4 yield exactly the same result.

5.4.3 Robot Reliability

Robot reliability may simply be described as the probability that a robot will perform its specified function satisfactorily for the stated time interval when used as per designed conditions. The general formula to obtain time-dependent robot reliability is expressed by [3,4]

$$
R_r(t) = \exp\left[-\int_0^t \lambda_r(t)\,dt \right] \tag{5.7}
$$

where

$R_r(t)$ is the robot reliability at time t.

$\lambda_r(t)$ is the robot's time-dependent failure rate (hazard rate).

Equation 5.7 can be used for obtaining the reliability function of a robot for any failure times probability distribution (e.g., Weibull, gamma, or exponential).

EXAMPLE 5.4

Assume that the time-dependent failure rate of a robot is expressed by

$$\lambda_r(t) = \frac{bt^{b-1}}{\gamma^{b-1}} \qquad (5.8)$$

where
b is the shape parameter.
t is time.
γ is the scale parameter.
$\lambda_r(t)$ is the hazard rate or time-dependent failure rate of the robot when its times to failure follows the Weibull distribution.

Obtain an expression for the robot reliability.
By substituting Equation 5.8 into Equation 5.7, we get

$$R_r(t) = \exp\left[-\int_0^t \frac{bt^{b-1}}{\gamma^{b-1}} dt \right]$$

$$= e^{-(t/\gamma)^b} \qquad (5.9)$$

Thus, Equation 5.9 is the expression for the robot reliability.

EXAMPLE 5.5

Assume that the constant failure rate of a robot is 0.0002 failures per hour. Calculate its reliability for a 15-h mission.
By substituting the given constant failure rate value of the robot into Equation 5.7, we obtain

$$R_r(t) = \exp\left[-\int_0^t (0.0002) dt \right]$$

$$= e^{-(0.0002)t} \qquad (5.10)$$

Inserting the given mission time value of the robot into Equation 5.10 yields

$$R_r(15) = e^{-(0.0002)(15)}$$

$$= 0.9970$$

Thus, the robot reliability for the specified mission period is 0.9970.

5.4.4 Robot's Hazard Rate

The robot's hazard rate or time-dependent failure rate is defined by [3,4]

$$\lambda_r(t) = -\frac{1}{R_r(t)} \cdot \frac{dR_r(t)}{dt} \tag{5.11}$$

where
  $R_r(t)$ is the robot reliability at time t.
  $\lambda_r(t)$ is the robot's hazard rate (time-dependent failure rate).

Equation 5.11 can be used for obtaining the robot's hazard rate when the robot's times to failure follows any time-continuous probability distribution (e.g., Weibull, gamma, and Rayleigh, exponential).

EXAMPLE 5.6

Assume that the reliability of a robot is expressed by

$$R_r(t) = e^{-\lambda_r t} \tag{5.12}$$

where
  $R_r(t)$ is the robot reliability at time t.
  λ_r is the robot's constant failure rate.

Obtain an expression for the robot's hazard rate.
  By substituting Equation 5.12 into Equation 5.11, we get

$$\lambda_r(t) = -\frac{1}{e^{-\lambda_r t}} \cdot \frac{de^{-\lambda_r t}}{dt}$$

$$= -\frac{1}{e^{-\lambda_r t}}[-\lambda_r e^{-\lambda_r t}]$$

$$= \lambda_r \tag{5.13}$$

Equation 5.13 is the expression for the robot's hazard rate, which is constant (i.e., it does not depend on time).

5.5 Robot Reliability Analysis Methods and Models for Performing Robot Reliability Studies

In the field of reliability engineering, there are many methods used to perform various types of reliability analyses. Some of these methods can be used quite effectively to perform robot reliability analysis. Four of these methods are as follows:

- *Failure modes and effect Analysis (FMEA).* This method was developed by the U.S. Department of Defense in the early 1950s and is considered an effective tool to perform analysis of each failure mode in the equipment/system for determining the effects of such failure modes on the entire equipment/system [12].

  The method is composed of six steps: (i) define equipment/system and its associated requirements, (ii) develop necessary ground rules, (iii) describe the equipment/system and all its associated functional blocks, (iv) identify all possible failure modes and their effects, (v) develop a list of critical items, and (vi) document the analysis [12–14]. Additional information on FMEA is available in Chapter 4 and in [4,15].

- *Fault tree analysis (FTA).* This method was developed at the Bell Telephone Laboratories in the early 1960s and is widely used to evaluate reliability of engineering systems during their design and development phase. A fault tree may simply be described as a logical representation of the relationship of primary/basic fault events that lead to the occurrence of a specified undesired event called the "top event."

  Additional information on fault tree analysis is available in Chapter 4 and in [4,16].

- *Markov method.* This method can be used in more cases than any other reliability evaluation method. It is concerned with modeling systems with constant failure and repair rates.

  Additional information on this method is available in Chapter 4 and in [4,17].

- *Parts count method.* This method is generally used during a bid proposal and early design phases to estimate the equipment/system failure rate. The method requires information on items such as quality levels of parts, quantities and types of generic parts, and the use environment of equipment/system/products.

  Additional information on the method is available in [4,18].

Over the years, many mathematical models have been developed to perform reliability-related studies of robots. Although the effectiveness of these models can vary significantly from one application area to another, some are being used quite successfully for performing reliability-related studies of robots in the industrial sector.

Many of these mathematical models were developed using the Markov method. Additional information about these models is available in Chapter 11 and in [3,11,19–21].

5.6 Reliability Analysis of Hydraulic and Electric Robots

As both hydraulic and electric robots are used in the industrial sector, this section presents reliability analyses of two typical hydraulic and electric robots by using the block diagram method [1–3]. Usually, for the purpose of design evaluation, it is assumed for both hydraulic and electric robots that all robot parts act in series (i.e., if any one part fails, the robot fails).

5.6.1 Reliability Analysis of the Hydraulic Robot

A hydraulic robot considered here contains five joints and, in turn, each joint is driven and controlled by a hydraulic servomechanism. The robot is subject to the following assumptions/factors [1,3]:

- Under high flow demand, an accumulator assists the pump to supply additional hydraulic fluid.
- The position transducer provides the joint angle codes and, in turn, the scanning of each code is carried out by a multiplexer.
- A conventional motor and pump assembly produces pressure.
- An operator makes use of a teach pendant for controlling the arm-motion in teach mode.
- Hydraulic fluid is pumped from the reservoir.
- The motion of each hydraulic actuator is controlled by a servo valve. This motion is transmitted directly or indirectly (i.e., through rods, chains, gears, etc.) to the specific limb of the robot and, in turn, each limb is coupled to a position transducer.
- An unloading valve is used for keeping pressure under the maximum limit.

The block diagram shown in Figure 5.1 represents the hydraulic robot under consideration with respect to reliability. Figure 5.1 shows that the hydraulic robot is composed of four subsystems (i.e., gripper subsystem, hydraulic pressure supply subsystem, drive subsystem, and electronic and

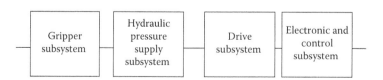

FIGURE 5.1
Block diagram of the hydraulic robot under consideration.

control subsystem) in series. In turn, as shown in the Figure 5.2 gripper subsystem (i.e., block diagram (a)) is composed of two parts (i.e., control signal and pneumatic system) in series, hydraulic pressure supply subsystem (i.e., block diagram (b)) is also composed of two parts (i.e., piping and hydraulic equipment) in series, and drive subsystem (i.e., block diagram (c)) is composed of five parts (i.e., joints 1, 2, 3, 4, and 5) in series.

With the aid of Figure 5.1, we have the following expression for the probability of the nonoccurrence of the hydraulic robot event (i.e., undesirable hydraulic robot arm movement causing damage to the robot/other equipment and possible harm to humans):

$$R_{hr} = R_1 R_2 R_3 R_4 \qquad (5.14)$$

where

R_{hr} is the hydraulic robot reliability or the probability of the nonoccurrence of the hydraulic robot event (i.e., undesirable robotic arm movement causing damage to the robot/other equipment and possible harm to humans).

R_1 is the reliability of the independent gripper subsystem.

R_2 is the reliability of the independent hydraulic pressure supply subsystem.

R_3 is the reliability of the independent drive subsystem.

R_4 is the reliability of the independent electronic and control subsystem.

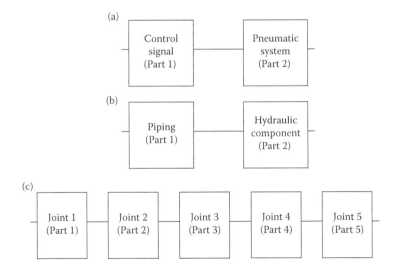

FIGURE 5.2
Block diagram representing three subsystems shown in Figure 5.1: (a) gripper subsystem, (b) hydraulic pressure supply subsystem, and (c) drive subsystem.

For independent parts, the reliabilities R_1, R_2, and R_3 of gripper subsystem, hydraulic pressure supply subsystem, and drive subsystem using Figure 5.2a–c, respectively, are

$$R_1 = R_{cs}R_{ps} \tag{5.15}$$

$$R_2 = R_p R_{hc} \tag{5.16}$$

and

$$R_3 = \prod_{i=1}^{5} R_{ji} \tag{5.17}$$

where
R_1 is the reliability of the gripper subsystem.
R_2 is the reliability of the hydraulic pressure supply subsystem.
R_3 is the reliability of the drive subsystem.
R_{cs} is the reliability of the control signal.
R_{ps} is the reliability of the pneumatic system.
R_p is the reliability of the piping.
R_{hc} is the reliability of the hydraulic component.
R_{ji} is the reliability of joint i, for $i = 1,2,3,4,5$.

For constant failure rates of independent subsystems shown in Figure 5.1, and, in turn, of their corresponding independent parts shown in Figure 5.2; from Equations 5.14 through 5.17, we obtain

$$R_{hr}(t) = e^{-\lambda_1 t} e^{-\lambda_2 t} e^{-\lambda_3 t} e^{-\lambda_4 t}$$
$$= e^{-\lambda_{cs} t} e^{-\lambda_{ps} t} e^{-\lambda_p t} e^{-\lambda_{hc} t} e^{-\sum_{i=1}^{5} \lambda_{it}} e^{-\lambda_4 t}$$
$$= e^{-(\lambda_{cs} + \lambda_{ps} + \lambda_p + \lambda_{hc} + \sum_{i=1}^{5} \lambda_{ji} + \lambda_4)t} \tag{5.18}$$

where
$R_{hr}(t)$ is the hydraulic robot reliability or the probability of the nonoccurrence of the hydraulic robot event (i.e., undesirable robotic arm movement, causing damage to the robot/other equipment and possible harm to humans) at time t.
λ_1 is the constant failure rate of the gripper subsystem.
λ_2 is the constant failure rate of the hydraulic pressure supply subsystem.
λ_3 is the constant failure rate of the drive subsystem.
λ_4 is the constant failure rate of the electronic and control subsystem.
λ_{cs} is the constant failure rate of the control signal (part 1).

λ_{ps} is the constant failure rate of the pneumatic system (part 2).
λ_p is the constant failure rate of the piping (part 1).
λ_{hc} is the constant failure rate of the hydraulic component (part 2).
λ_{ji} is the constant failure rate of the joint i, for $i = 1,2,3,4,5$.

By integrating Equation 5.18 over the time interval $[0,\infty]$, we get

$$MTTOHRE = \int_0^\infty e^{-(\lambda_{cs}+\lambda_{ps}+\lambda_p+\lambda_{hc}+\Sigma_{i=1}^5\lambda_{ji}+\lambda_4)t}\,dt$$

$$= \frac{1}{\lambda_{cs} + \lambda_{ps} + \lambda_p + \lambda_{hc} + \Sigma_{i=1}^5 \lambda_{ji} + \lambda_4} \tag{5.19}$$

where
MTTOHRE is the mean time to the occurrence of the hydraulic robot event: undesirable arm movement causing damage to the robot/other equipment and possible harm to humans.

EXAMPLE 5.7

Assume that the following reliability data values are given for the above type of hydraulic robot:

$$R_{j1} = R_{j2} = R_{j3} = R_{j4} = R_{j5} = 0.92, \quad R_4 = 0.95,$$
$$R_{cs} = 0.86, \quad R_{ps} = 0.96, \quad R_p = 0.94, \quad \text{and} \quad R_{hc} = 0.9$$

Calculate the non occurrence probability of the event: undesirable hydraulic robot arm movement causing damage to the robot/equipment and possible harm to humans.

By substituting the given data values into Equations 5.15 through 5.17, we get

$$R_1 = (0.86)(0.96) = 0.8256$$
$$R_2 = (0.94)(0.9) = 0.846$$

and

$$R_3 = (0.92)(0.92)(0.092)(0.92)(0.92) = 0.6591$$

By substituting the above-calculated values and the given data value into Equation 5.14, we get

$$R_{hr} = (0.8256)(0.846)(0.6591)(0.95)$$
$$= 0.4373$$

Thus, the non occurrence probability of the event: undesirable hydraulic robot arm movement causing damage to the robot/equipment and possible harm to humans, is 0.4373.

EXAMPLE 5.8

Assume that the specified constant failure rate values of the above type of hydraulic robot are $\lambda_{cs} = 0.0007$ failures per hour, $\lambda_{ps} = 0.0006$ failures per hour, $\lambda_p = 0.0005$ failures per hour, $\lambda_{htc} = 0.0004$ failures per hour, $\lambda_{j1} = \lambda_{j2} = \lambda_{j3} = \lambda_{j4} = \lambda_{j5} = 0.0003$ failures per hour, and $\lambda_4 = 0.0002$ failures per hour, respectively.

Using Equation 5.19, calculate the mean time to the occurrence of the hydraulic robot event: undesirable arm movement causing damage to the robot/other equipment and possible harm to humans.

By substituting the specified data values into Equation 5.19, we get

$$MTTOHRE = \frac{1}{0.0007 + 0.0006 + 0.0005 + 0.0004 + 5(0.0003) + 0.0002}$$
$$= 256.41 \text{ h}$$

Thus, the mean time to the occurrence of the hydraulic robot event: undesirable robot arm movement causing damage to the robot/other equipment and possible harm to humans is 256.41 h.

5.6.2 Reliability Analysis of the Electric Robot

An electric robot considered here is one that performs a "normal" industrial task, while its programming and maintenance are carried out by humans. The robot is subject to the following assumptions/factors [2,3]:

- The motor shaft rotation is transmitted to the appropriate limb of the robot through a transmission unit.
- A microprocessor control card controls each and every joint.
- A supervising computer/controller directs all joints.
- A direct current (DC) motor actuates each joint.
- An interface bus allows interaction between the supervisory controller and the joint control processors.
- The transducer sends all appropriate signals to the joint controller.
- Each and every joint is coupled with a feedback transducer (encoder).

The block diagram shown in Figure 5.3 represents the electric robot under consideration with respect to reliability. Figure 5.3 shows that the electric robot is composed of two hypothetical subsystems (i.e., no movement due to external factors (subsystem 1) and no failure within the robot causing its

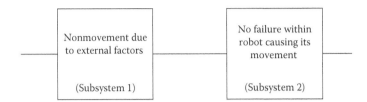

FIGURE 5.3
Block diagram for estimating the nonoccurrence probability (reliability) of the undesirable movement of the electric robot.

movement (subsystem 2) in series. In turn, as shown in the Figure 5.4, subsystem 1 (i.e., block diagram (a)) is composed of two hypothetical elements (i.e., element A and element B) in series and subsystem 2 (i.e., block diagram (c)) is composed of five parts (i.e., interface, end effector, supervisory computer/controller, joint control, and drive transmission) in series.

Furthermore, as shown in Figure 5.4, element A (i.e., block diagram (b)) is composed of two hypothetical subelements (i.e., subelement X and subelement Y) in series.

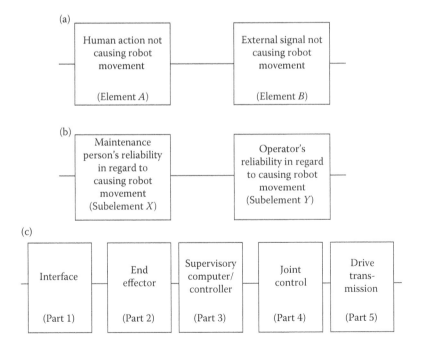

FIGURE 5.4
Block diagram representing two subsystems shown in Figure 5.3: (a) subsystem 1, (b) element A of subsystem 1, and (c) subsystem 2.

With the aid of Figure 5.3, we have the following expression for the probability of nonoccurrence of the undesirable electric robot movement (reliability):

$$R_{er} = R_1 R_2 \qquad (5.20)$$

where

R_{er} is the probability of nonoccurrence (reliability) of the undesirable electric robot movement.
R_1 is the reliability of the independent subsystem 1.
R_2 is the reliability of the independent subsystem 2.

For independent elements, the reliability of subsystem 1 in Figure 5.4a is

$$R_1 = R_A R_B \qquad (5.21)$$

where

R_A is the reliability of element A.
R_B is the reliability of element B.

For independent and hypothetical subelements, the reliability of element A in Figure 5.4b is

$$R_A = R_X R_Y \qquad (5.22)$$

where

R_X is the maintenance person's reliability with regard to causing the robot's movement (subelement X).
R_Y is the operator's reliability with regard to causing the robot's movement (subelement Y).

Similarly, for independent parts, the reliability of subsystem 2 in Figure 5.4c is

$$R_2 = R_i R_e R_s R_j R_d \qquad (5.23)$$

where

R_i is the reliability of the interface.
R_e is the reliability of the end-effector.
R_s is the reliability of the supervisory computer/controller.
R_j is the reliability of the joint control.
R_d is the reliability of the drive transmission.

EXAMPLE 5.9

Assume that the following reliability data values are given for an above type of electric robot:

$$R_B = 0.96, \ R_X = 0.91, \ R_Y = 0.92, \ R_i = 0.90,$$
$$R_e = 0.93, \ R_s = 0.94, \ R_j = 0.95, \quad \text{and} \quad R_d = 0.97$$

Calculate the probability of nonoccurrence (reliability) of the undesirable electric robot movement.

By substituting the specified data values into Equations 5.22 and 5.23, we get

$$R_A = (0.91)(0.92) = 0.8372$$

and

$$R_2 = (0.90)(0.93)(0.94)(0.95)(0.97) = 0.7250$$

By inserting the above-calculated value for R_A and the given data value for R_B into Equation 5.21, we get

$$R_1 = (0.8372)(0.96) = 0.8037$$

By substituting the above-calculated values into Equation 5.20, we obtain

$$R_{er} = (0.8037)(0.7250) = 0.5827$$

Thus, the probability of nonoccurrence (reliability) of the undesirable electric robot movement is 0.5827.

5.7 Problems

1. Discuss classifications of robot failures and their causes.
2. What are the dictating factors for robot effectiveness?
3. Write down the formula to estimate mean time to robot-related problems.
4. Assume that the annual production hours of a robot and its annual downtime due to failures are 5000 and 150 h, respectively. During that period the robot failed four times. Calculate the mean time to robot failure.

5. Assume that the constant failure rate, λ_r, of a robot is 0.0004 failures per hour and its reliability is expressed by

$$R_r(t) = e^{-\lambda_r t} \qquad\qquad (5.24)$$
$$= e^{-(0.0004)t}$$

where

$R_r(t)$ is the robot reliability at time t.

Calculate the mean time to robot failure with the aid of Equations 5.3 and 5.4 and comment on the end result.

6. Prove Equation 5.9 by using Equations 5.7 and 5.8.

7. Assume that the reliability of a robot is expressed by Equation 5.9. Obtain an expression for the robot's hazard rate.

8. Discuss at least four methods that can be used to analyze robot reliability.

9. Compare a hydraulic robot with an electric robot with respect to reliability.

10. Discuss with results of the robot reliability surveys concerning 37 robots of four different designs used in three different companies, covering almost 22,000 robot production hours.

References

1. Khodanbandehloo, K., Duggan, F., Husband, T.F., Reliability assessment of industrial robots, *Proceedings of the 14th International Symposium on Industrial Robots*, 1984, pp. 209–220.
2. Khodabandehloo, K., Duggan, F., Husband, T.F., Reliability of industrial robots: A safety viewpoint, *Proceedings of the 7th British Robot Association Annual Conference*, 1984, pp. 233–242.
3. Dhillon, B.S., *Robot Reliability and Safety*, Springer-Verlag, New York, 1991.
4. Dhillon, B.S., *Design Reliability: Fundamentals and Applications*, CRC Press, Boca Raton, Florida, 1999.
5. Dhillon, B.S., *Human Reliability: With Human Factors*, Pergamon Press, New York, 1986.
6. Dhillon, B.S., *Reliability Engineering in Systems Design and Operation*, Van Nostrand Reinhold Company, New York, 1983.
7. Dhillon, B.S., *Reliability in Computer Systems Design*, Ablex Publishing, Norwood, New Jersey, 1987.
8. Herrmann, D.S., *Software Safety and Reliability*, IEEE Computer Society Press, Los Alamitos, California, 1999.
9. Young, J.F., *Robotics*, Butterworth, London, 1973.

10. Jones, R., Dawson, S., People and robots: Their safety and reliability, *Proceedings of the 7th British Robot Association Annual Conference*, 1984, pp. 243–258.
11. Dhillon, B.S., *Applied Reliability and Quality: Fundamentals, Methods, and Procedures*, Springer, London, 2007.
12. Omdahl, T.P., Editor, *Reliability, Availability, and Maintainability (RAM) Dictionary*, American Society for Quality Control (ASQC) Press, Milwaukee, Wisconsin, 1988.
13. Coutinho, J.S., Failure effect analysis, *Transactions of the New York Academy of Sciences*, Vol. 26, Series II, 1963–1964, pp. 564–584.
14. Bureau of Naval Weapons, Department of the Navy, *General Specification for Design, Installation, and Test of Aircraft Flight Control Systems, MIL-F-18372 (Aer)*, Washington, DC, Para. 3.5.2.3.
15. Palady, P., *Failure Modes and Effects Analysis*, PT Publications, West Palm Beach, Florida, 1995.
16. U.S. Nuclear Regulatory Commission, *Fault-Tree Handbook, Report No. NUREG-0492*, Washington, DC, 1981.
17. Shooman, M.L., *Probabilistic Reliability: An Engineering Approach*, McGraw-Hill Book Company, New York, 1968.
18. US Department of Defense, *Reliability Prediction of Electronic Equipment, MIL-HDBK-217*, Washington, DC.
19. Dhillon, B.S., Fashandi, A.R.M., Robotic systems probabilistic analysis, *Microelectronics and Reliability*, Vol. 37, 1997, pp. 211–224.
20. Dhillon, B.S., Fashnadi, A.R.M., Stochastic analysis of a robot machine with duplicate safety units, *Journal of Quality in Maintenance Engineering*, Vol. 5, No. 2, 1999, pp. 114–127.
21. Dhillon, B.S., Li, Z., Stochastic analysis of a system with redundant robots, one built-in safety unit, and common cause failures, *Journal of Intelligent and Robotics Systems*, Vol. 45, 2006, pp. 137–155.

6

Robot Safety

6.1 Introduction

Today, robots are being used in many diverse areas and applications, and their safety-related problems have increased quite significantly. Each area and application may call for specific precautions for programmers, operators, robot systems, maintenance workers, and so on. In order to meet needs such as these, various organizations have, over the years, developed documents that are specifically concerned with robot safety.

Two examples of such documents are the *American National Standard for Industrial Robots and Robot Systems—Safety Requirements* [1] and the Japanese Industrial Safety and Health Association document titled, "An Interpretation of the Technical Guidance on Safety Standards in the Use, Etc., of Industrial Robots" [2].

Generally, factors such as replacing humans in performing hazardous and difficult tasks and increasing productivity play an important role in the use of robots in the industrial sector. Improper consideration to safety in robotics planning may create hazardous conditions other than those that the robot may be replacing. It simply means that for successful applications of robots, safety must be considered with utmost care during the planning phase.

This chapter presents various important aspects of robot safety.

6.2 Robot Safety: Problems and Hazards

Safety professionals concerned with robots face many unique robot-related safety problems. Some of these problems are as follows [3,4]:

- Robots generate potentially hazardous conditions because they manipulate items of varying weights and sizes.
- The presence of a robot receives great attention from humans in the surrounding area, who are often quite ignorant of the possible associated hazards.

- Robot-related maintenance procedures may lead to hazardous situations.
- A robot may lead to a high risk of fire or an explosion if it is placed in an unfriendly environment.
- In the event of the occurrence of a control, hydraulic, or a mechanical failure, robots may move out of their programmed area zones and strike something or they may throw some small item.
- Normally, robots function quite closely with other machines and humans in the same environment. In particular, humans are subject to collisions with the moving parts of robots, tripping over loose power/control cables (if any), and of being pinned down.
- Various safety-associated electrical design problems can occur in robots. Some examples of these problems are poorly designed power sources, fire hazards, and potential for electric shock.
- Robot mechanical design-related problems may result in hazards such as grabbing, pinning, and pinching.
- Management attitudes very much lead to miscomprehension of robot safety-related concepts.

Three basic types of robot hazards are shown in Figure 6.1 [3–5]—trapping, impact, and the hazards that develop from the application.

The trapping hazard is usually the result of robot movements in regard to fixed objects such as pests and machines in the same area. The movements of the auxiliary equipment could be the other possible causes. Two examples of such equipment are pallets and carriages.

The impact hazard involves being struck by the moving part of a robot or by parts or items being carried by the robot. This hazard also includes being struck by flying objects that are ejected or dropped by the robot.

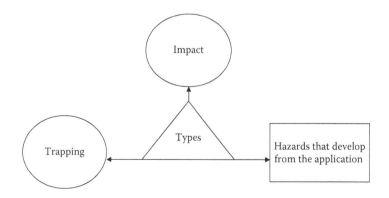

FIGURE 6.1
Basic types of robot-related hazards.

Some of the hazards that develop from the application are burns, electric shocks, exposure to toxic substances, and arc flash. The most prevalent causes of these hazards are as follows [6]:

- Control errors
- Mechanical-related problems
- Human error
- Unauthorized access

Control errors may simply be described as intrinsic faults within the robot's control system. Some examples of control errors are electrical interference, faults in the hydraulic, pneumatic, or electrical subcontrol systems, and software problems. The mechanical-related problems arise from the application or the robot system itself. Generally, mechanical hazards occur during the transfer of items with sharp edges, and from malfunctions due to poor maintenance or overloading.

Human error occurs when humans enter the robot's protected work zone and carry out their required tasks close to it. Usually, complacency or lack of care during human interaction leads to human error. Finally, unauthorized access is probably the most prevalent source of hazard; it can easily be controlled by developing proper standards for access, and then following these effectively.

6.3 Roles of Robot Manufacturers and Users in Robot Safety

As both robot manufacturers and users profit from robot-related safety, each has a very important role to play in its effective promotion. The basic responsibility of robot manufacturers is to design and manufacture safe robots. The design-related safety measures of robots can be grouped under the following five categories [6–8]:

- *Mechanical subsystem.* Safety in mechanical subsystems of robots can be built in the design process, by taking into consideration the sources of hazards and appropriate measures to eliminate them. Kinetic energy storage capacity, poor reliability, incompatibility of materials with the work environment, and pinch-points are the main sources of hazard in the mechanical subsystem. Furthermore, during the design process, proper attention must be paid to specific sources of danger such as uncovered ball screws, exposed motors, protruding linkages, and flopping cables and hoses.

The robot kinetic energy level can be minimized by optimizing the design for minimum arm inertia. The mechanisms such as dynamic braking, dumping within the drive system, mechanical stops and shock absorbers, and conversion to elastic energy in spring-type parts can be used to attain control of robotic kinetic energy.

- *Electrical/electronic subsystem.* There are various safety-related features that can be designed into the robot's electrical/electronic subsystem, including emergency stops that can be activated by an operator in the event of a hazardous situation, intrinsic safety circuits, input/output signal conditioning, and equipping the teach pendant with a dead-man switch.

- *Operational procedures.* These procedures are an integral part of robot design and they must be provided effectively for the safe operation of the robot. They include items such as operating manuals, sets of instructions, and appropriate precautions.

- *Control subsystem's algorithm.* Robot-related safety can be improved appreciably through the effective control subsystem algorithm design. Therefore, proper attention has to be paid to items such as the regulation and control of the maximum operating speed of the drives and the definition of the tolerable feedback system's following error.

- *Operational and control subsystems' software.* The most safety-related awareness related to the robot can be introduced in the design of the software for the operational and control subsystem. It is to be noted that the capabilities that can be designed into robots are only useful if the logic system is active as well as the robots themselves are calibrated properly. Some examples of the capabilities are
  - Imposition of a maximum robot-operating speed during teaching
  - A response system to switch-off signals from outside interfaces
  - A response system to communication data-flow abnormalities
  - An emergency response system to an unusual velocity on every server-controlled drive
  - A response to interrupts generated by limits associated with hardware travel

Past experiences indicate that the users of robots play an equally important role with regard to robot-related safety. The robot users' safety-related responsibilities may be grouped in the following three categories [7]:

- *The engineering/maintenance department.* The responsibilities of the engineering/maintenance department include installation according to the instructions of the manufacturer, locating all control stations, with the exception of the Pendant Control (whenever required) outside the restricted operating zone; a proper maintenance program, and effective training of maintenance workers.

- *The management.* With regard to safety, the basic two responsibilities of the robot user's management team are in appreciating the safety implications of robotization and supporting the safety department/ unit to promote the special features of robot safety.

- *The safety department.* The safety department has many responsibilities, including installing appropriate warning signs or other measures in the robot work envelope, installing barriers with interlocking gates at the work boundary, providing appropriate safety-related training to all concerned personnel, collaborating with other groups concerned with the purchase and the use of robots, and keeping abreast with the latest developments associated with robot-related safety.

6.4 Safety Considerations in Robot Design, Installation, Programming, and Operation and Maintenance Phases[*]

In order to minimize the problems of robot-related safety, it is absolutely essential to carefully consider the safety factor during the robot design, installation, programming, and operation and maintenance phases. Some useful safety-related guidelines concerning each of these four phases are presented below, separately [2,8–11].

6.4.1 Robot Design Phase

The robot design safety features may be categorized under the following three classifications:

- *Mechanical.* The mechanical safety features include eliminating sharp corners, designing teach pendant ergonomically, ensuring the existence of mechanisms for releasing the stopped energy, having drive mechanism covers, putting guards on items such as belts, gears, and pulleys, having dynamic brakes for software crash or power failure, and having several emergency stop buttons.

---

[*] It is to be noted that the material presented in this section may overlap to a certain degree with the preceding section, but is included here for completeness.

- *Electrical.* The electrical safety features include ensuring the internal safety of the robot so that it will not ignite in a combustible environment, having built-in hose and cable routes using adequate insulation, sectioning and providing panel covers, designing wire circuitry capable of stopping the robot's movement and locking its brakes, minimizing the effects of electromagnetic and radiofrequency interferences, having a fuse "blow" long before human crushing pressure is experienced, and eliminating the risk of an electric shock.
- *Software.* The software safety features include using a procedure of checks to determine why a failure occurred, having built-in commands, periodically examining the built-in self-checking software for safety, having a standby power source for the robot's functioning with programs in random access memory, prohibiting a restart by merely resetting a single switch, providing a robot motion simulator, and having a restart approach after experiencing an emergency stop.

6.4.2 Robot Installation Phase

There are many safety features that related to the robot installation phase. Some of these features are as follows:

- Installing electrical cables according to electrical codes
- Installing appropriate interlocks, sensing devices, and so on
- Providing an appropriate level of illumination to humans concerned with the robot
- Placing robot controls outside the hazard area
- Identifying the danger zones with the aid of codes, signs, line markings, and so on
- Ensuring the accessibility and visibility of emergency stops
- Placing an appropriate shield between the robot and humans
- Providing protection to control circuitry by filtering spikes and surges
- Using vibration-reducing pads when required
- Adding pads, cushions, and so on to possible collision points with humans
- Controlling environmental factors as necessary
- Labeling the sources of stored energy
- Distancing circuit boards from electromagnetic fields
- Installing the required interlocks to interrupt robot motion

6.4.3 Robot Programming Phase

Safety during the robot programming phase is as important as during its design and installation phases. As per [10], a study of 131 cases reported that programmers/setters were at the highest risk, accounting for 57% of the accidents; the comparative figures for fault clearance personnel, maintenance personnel (repair/servicing), and operators were 26%, 4%, and 13%, respectively.

There are various factors that characterize a robot programmer's work conditions, including working in a bending position (i.e., normally in the robot's movement zone), frequently changing position because the torch's tip is concealed by the clamping device, and being subject to stress by improper lighting. Some of the safety measures that can be taken into consideration with respect to robot-related safety are as follows [10]:

- Pressure-sensitive mats on the floor at the position of the programmer
- A manual programming device containing an emergency off switch
- Designing the work area of the programmer in such a way that it eliminates unnecessary stress
- Hold-to-run-buttons
- Mandatory reduced speed
- Marking the programming positions
- Planning the programmer's location outside the movement zone (i.e., in a space facing the robot, semiraised on a platform, etc.)
- Turning off safety-related devices with a key only
- Locked turntables during programming

6.4.4 Robot Operation and Maintenance Phase

There are many safety-related measures associated with the robotic operation and maintenance phases [4,8,11]. Some of these measures require preventive maintenance regularly and use only the approved parts, develop the necessary safety operations and maintenance procedures, ensure that only authorized and trained personnel operate and maintain robots, report, investigate, and repair any faults or unusual robot motions promptly, ensure the operational readiness of all safety devices (e.g., barriers, interlocks, and guards), block out all concerned power sources during maintenance, observe all government codes and other regulations concerned with the operation and maintenance of robots, provide the initial and periodic training to people associated with robotic operation and maintenance, and make certain that all emergency stops are functional.

6.5 Robot-Related Safety Problems Causing Weak Points in Planning, Design, and Operation

There are many weak points in planning, design, and operation, which result in robot-related safety problems in an industrial setting [1,6]. Some of the weak points concerned with planning are as follows:

- *Poor organization of work.* This is an important factor, particularly in programming and stoppages.
- *Improper safety devices.* These comprise improper guards (i.e., too low, containing gaps, or being close to hazard points) and faulty emergency shutdown circuits.
- *Poor spatial arrangement.* This can lead to confusion and the possibility of collision.
- *Unsafe or confused linkages.* These are basically concerned with interfaces between individual machines.

Some of the weak points in design with respect to robot-related safety in industries are shown in Figure 6.2 [12].

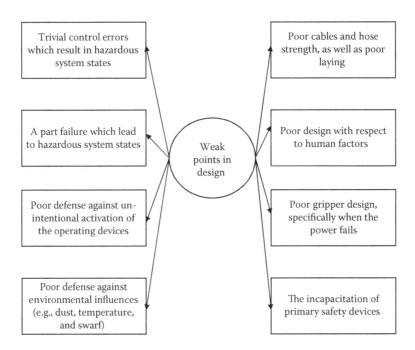

FIGURE 6.2
Weak points in design with respect to robot safety.

Finally, some of the weak points in operational procedure with respect to robot-related safety in industrial settings are as follows:

- Permitting counter-safety working procedures during a stoppage
- Poor training to workers who are directly or indirectly concerned with industrial robots
- Failure to give feedback to personnel involved in design and layout regarding weak spots and how to ensure their removal

6.6 Robot Safeguard Approaches

There are many robot safeguard approaches [5]. Some of these approaches are described below, separately [5,6].

6.6.1 Warning Signs

The warning signs are used in conditions where robots, by virtue of their size, speed, and inability to impart significant amount of force, cannot injure humans. Two examples of such conditions are laboratory and small-part assembly robots. Robots such as these require no special safeguarding as warning signs are quite sufficient for the uninformed humans in the area.

However, it is to be noted that warning signs are quite useful for all robot application areas, irrespective of whether robots possess the ability to injure humans or not.

6.6.2 Physical Barriers

Physical barriers are quite useful to safeguard humans in the work areas of robots. However, they are not the ultimate solution to a robot-related safety problem in many cases. The objective of such barriers is to stop humans from reaching over, under, around, or through the barriers into the prohibited robot work zone [1]. Some examples of physical barriers are chain-link fences, safety rails, plastic safety chains, and tagged-rope barriers. Some useful guidelines concerning physical barriers are the following [6]:

- Use safety rails in conditions where projectiles are clearly not a problem at all
- Avoid trapping points within the framework of a barrier by providing a sufficient buffer space between the barrier and the work area
- Chain-link fences and safety rails are very useful in circumstances where intrusion is considered a particular problem

- Make use of fences in circumstances where long-range projectiles are considered a hazard

In any case, when the installation of a peripheral physical barrier is considered, the following questions have to be asked clearly [13]:

- How were perimeter dimensions established?
- Are the perimeter dimensions reliable?
- Can the perimeter be bypassed easily?
- What is being protected?
- Is the protection effective?

6.6.3 Intelligent Systems

Intelligent systems make their decisions through remote sensing, hardware, and software. In order to achieve an effective intelligent collision-avoidance system, the robot's operating environment has to be restricted and special sensors and software has to be used widely. This calls for the need for a sophisticated computer for making the right decisions and real-time computations.

Finally, it is to be noted that in most industrial settings, it is generally not possible to restrict the environment.

6.6.4 Flashing Lights

Flashing lights are another robot safeguard approach; this approach calls for the installation of a flashing light on the robot itself or at the perimeter of the robot work area. The objective of the flashing light is to alert people in the area that programmed motion is happening or could happen any time.

It is to be noted with care that whenever the flashing lights approach is employed, ensure that the flashing lights approach is employed, ensure that the flashing light is energized continuously during the period when the robot drive power is turned on.

6.7 Common Robot Safety Features and Their Functions

A well-designed robot incorporates, to the greatest extent possible, safety features that clearly take into consideration all modes of robot operation (i.e., normal operation, programming, and maintenance) [14,15]. Nonetheless, some of the safety features are generally common to all robots and the others are specific to the robot types. Some of the common robot safety features, along with their corresponding intended functions, given in parentheses, are as follows [15]:

- Power-on button (it energizes all machine power)
- Stop button (it removes control and manipulator power)
- Power disconnect switch (it removes all power at the machine junction box)
- Slow-speed control (it permits program execution at reduced speeds)
- Arm-power-only button (it applies power to the manipulator only)
- Teach pendant trigger (it must be held by the operator for arm power in teach mode)
- Line indicator (it indicates that incoming power is connected at the junction box)
- Hydraulic fuse (it protects against high-speed motion/force in teach mode)
- Manual/automatic dump (it provides means to relieve hydraulic/pneumatic pressure)
- Step button (it permits program execution one step at a time)
- Condition indicators and messages (these provide visual indication by lights or display screens of the condition of the system)
- Remote connections (these permit remote control of essential machine/safety functions)
- Parity checks, error detecting, and so on (whereby the computer approaches for self-checking a variety of functions)
- Hardware stops (absolute control on travel/movement limits)
- Program reset (this drops the system out of playback mode)
- Hold/run button (it stops arm motion, but leaves power on)
- Control-power-only button (it applies power to the control section only)
- Software stops (computer-controlled travel limit)
- Teach/playback mode selector (it provides the operator with control over the operating mode)
- Servo-motor brake (it maintains the arm position at a standstill)

6.8 Safety Considerations for Robotized Welding Operations

Today, in industries, robots are often used to perform various types of welding operations. Welding robots undertake welding processes such as gas tungsten arc, resistance (spot), gas metal arc, and laser [6,16].

There are many factors that have to be considered with care for the safe operation of welding robots. Twelve of these factors are shown in Figure 6.3 [6,16]. The factor "torch motion simulation" is concerned with the simulation of the welding torch motion prior to striking the arc. The factor "common ground" calls for not sharing the common ground in situations where a welding robot is joining conventional welding systems with high-frequency power supplies.

The factor "area contamination" is concerned with ensuring that welding robots do not contaminate areas where spatter or harmful light rays may generate adverse effects.

The factor "environment" is concerned with finding an area that is free from dust, high humidity, vibration, oil, and smoke and also has low traffic.

The factor "power switches" is concerned with locking such switches to the control panel and welding power supply. The factor "interlocking switches" is concerned with installing such switches to the welding jigs and fixtures, which indicate safe clamping.

The factor "robot floor" is concerned with choosing a suitable floor, surrounding and directly beneath the robot. In this case, it is to be noted that

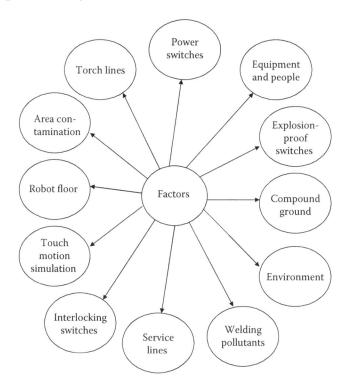

FIGURE 6.3
Factors to be considered for the safe operation of welder robots.

metal plates and explosion-proof concrete offer a good solution for problems caused by sparks and hot metal falling to the floor. The factor "equipment and people" is concerned with preventing damage and injury to equipment and people, respectively. In this case, it is to be noted that by placing spatter shields and ultraviolet light-filtering screens around the welding table, the positioner and workstation can prevent damage and injury to equipment and people.

The factor "service lines" includes electricity, water, air, exhaust, and ventilation lines. Here, the aim should be to locate such service lines underneath the floor or in floor channels away from the welding torch and torch lines, in addition to locating them away from the robot's work area. Furthermore, avoid having water and air lines in the same channel. The factor "torch lines" is concerned with avoiding the entanglement of such lines with the joints of the robot during extreme movement. The items such as pivots, strings, pulleys, and counterweights can be used to eliminate this problem.

The factor "welding pollutants" is concerned with alleviating problems of welding pollutants. In this case, an electrostatic precipitator may help to reduce pollutants as well as the use of a centrally located charcoal collector to reduce welding gases. It is to be noted that this collector draws the residual gases through its torch attachment. Thus, it is very useful in circumstances where it is not possible to have an underfloor exhaust system as well as where the robot has to perform many welds over fairly long distances.

Finally, the factor "explosion-proof switches" is concerned with using such switches whenever possible.

As there are specific hazards and safeguards associated with robotized laser welding, gas-shielded arc welding, and resistance welding, each of these topics is discussed below, separately [6,16].

6.8.1 Robotized Laser Welding Hazards and Safety Measures

Laser welding is the type of welding where energy is generated by a laser beam and focused at a point. The hazards that emanate from the use of robotized laser welding are those connected with fumes and electromagnetic radiation. Welding lasers can cause fire and hazards to both eyes and skin (diffuse reflections).

When the robot is in its normal operation mode, there are many safety measures associated with power lasers. Some of these safety measures are as follows [6,16]:

- *Safety measure I: Emission indicator.* It indicates that the laser is in its active mode.
- *Safety measure II: Beam attenuator.* It helps to stop light from the laser entering the enclosure while humans are still inside.

- *Safety measure III: Enclosure.* It helps to stop exposure to laser light.
- *Safety measure IV: Remote interlock.* It is used for stopping access to the power laser enclosure during laser action.
- *Safety measure V: Warning signs.* These are to be placed at accesses to the enclosure for their effectiveness.
- *Safety measure VI: Key control.* It is useful for stopping unauthorized use.
- *Safety measure VII: Training.* It is very helpful for familiarization with the product and for hazard control.

6.8.2 Hazards and Safety Measures for Robotized Gas-Shielded Arc Welding

Gas-shielded arc welding is the type of welding where both the arc and molten pool are shielded from the atmosphere by a gas [16]. When this type of welding is robotized, there are many types of associated hazards. Some of these hazards are resulting fumes, hot metal sparks, electric shock from the torch, fire from hot metal welded parts, electromagnetic radiation, mechanical hazards, and program corruption [6,16]. For modes normal, programming, and maintenance solutions to these hazards are available in [16].

For the normal mode of robotized operations, selected hazards associated with gas-shielded arc welding and their corresponding safety measures (given in parentheses) are as follows [6]:

- Phosgene from chlorinated solvents on work items (dry all parts with care after any cleaning process before welding)
- Ozone from arc (provide appropriate fume extraction fans in the general area of work)
- Electric shock from torch (place torch inside the work perimeter)
- Fire from hot metal sparks (ensure that no flammable materials exist inside the perimeter guard and that all cables and pipes within the perimeter guard framework are properly protected)
- Fire from hot metal welded parts (ensure that the unloaded items are properly kept away from flammable materials)
- Sparks from hot metal (position an interior perimeter guard and carefully ensure that there is no flammable material inside the guarded workspace. Also, keep all pipes and cables inside the guarded workspace properly protected)
- Smoke from leftover oil on work items (minimize oil on the work surface to decrease oil fumes as well as to ensure good welds)
- Magnetic radiation (minimize this radiation to a normal level; computer enclosures should furnish an adequate shield to electronics)

6.8.3 Hazards and Safety Measures for Robotized Resistance Welding

Resistance welding is the type of welding in which, at some point during the process, force is put to the surfaces in contact, and in which welding heat is produced by sending electric current through the resistance at, and adjacent to, these surfaces [6,16]. The use of robots is quite useful to minimize some of the hazards that are traditionally associated with resistance welding.

These associated hazards may be categorized into the following three groups [6]:

- *Electrical hazards.* These hazards are magnetic fields, high/medium voltage, and access to welding control equipment. The safety measures applicable to the normal mode of robot operation are the identification of workers wearing pacemakers, placing primary supply and welding transformers in a safe location, and locking to stop unauthorized access.
- *Mechanical hazards.* Trapping by work-handling equipment and shocks from low-voltage conductors are the two examples of mechanical hazards. The safety measures applicable to the normal mode of robotized operation are providing satisfactory guarding/ interlocks to stop access to the hazard zone and balance support for minimizing mechanical stress on the robot arm.
- *Other hazards.* These hazards are fumes, noise, and fire. Some of the safety measures applicable to the normal mode of robotized operation are placing extractor fans in an "on" position, keeping closed containers of weld sealers, and silencers on the air exhaust.

6.9 Problems

1. What are the unique robot safety problems? List at least six such problems.
2. What are the basic robot hazards? Discuss each of these hazards in detail.
3. Discuss the roles of robot manufacturers and users with regard to safety.
4. Discuss safety considerations in robot design and installation phases.
5. Discuss safety considerations in robot programming and operations and maintenance phases.
6. List at least eight weak points in design with respect to robot-related safety.

7. Describe the following two robot safeguard approaches:
   a. Warning signs
   b. Flashing lights

8. List at least 10 common robot-related safety features along with their functions.

9. What are the factors to be considered for the safe operation of welder robots?

10. What are the robotized laser welding hazards and safety measures?

References

1. American National Standards Institute, *American National Standard for Industrial Robots and Robot Systems—Safety Requirements, ANSI/RIA R15.06-1986*, New York, 1986.

2. Japanese Industrial Safety and Health Association, *An Interpretation of the Technical Guidance on Safety Standards in the Use, Etc., of Industrial Robots*, Tokyo, 1985.

3. Ziskovsky, J.P., Risk analysis and the R^3 factor, *Proceedings of the Robots 8th Conference*, Vol. 2, June 1984, pp. 15.9–15.21.

4. Ziskovsky, J.P., Working safely with industrial robots, *Plant Engineering*, May 1984, pp. 81–85.

5. Addison, J.H., *Robotic Safety Systems and Methods: Savannah River Site, Report No. DPST-84-907 (DE 35-008261)*, December 1984, issued by E.I. du Pont de Nemours and Company, Savannah River Laboratory, Aiken, South Carolina 29808.

6. Dhillon, B.S., *Robot Reliability and Safety*, Springer-Verlag, New York, 1991.

7. Ramachandran, V., Vajpayee, S., Safety in robotic installations, *Robotics and Computer-Integrated Manufacturing*, Vol. 3, 1987, pp. 301–309.

8. Russell, J.W., *Robot Safety Considerations: A Checklist, Professional Safety*, December 1983, pp. 36–37.

9. Bellino, J.P., Meagher, J., Design for safeguarding, *Proceedings of the Robots East Seminar*, Boston, Massachusetts, October 1985, pp. 24–37.

10. Nicolaisen, P., Ways of improving industrial safety for the programming of industrial robots, *Proceedings of the 3rd International Conference on Human Factors in Manufacturing*, November 1986, pp. 263–276.

11. Jiang, B.C., Robot safety: Users' guidelines, *Trends in Ergonomics/Human Factors III*, ed. W. Karwowski, Elsevier, Amsterdam, 1986, pp. 1041–1049.

12. Akeel, H.A., Intrinsic robot safety, *Working Safely with Industrial Robots*, ed. P.M. Strubhar, Robotics International of the Society of Manufacturing Engineers, Publications Development Department, One SME Drive, P.O. Box 930, Dearborn, Michigan, 1986, pp. 61–68.

13. Marton, T., Pulaski, J.L., Assessment and development of HF related safety designs for industrial robots and robotic systems, *Proceedings of the Human Factors Society 31st Annual Meeting*, 1987, pp. 176–180.

14. Bararett, R.J., Bell, R., Hudson, P.H., Planning for robot installation and maintenance: A safety framework, *Proceedings of the 4th British Robot Association Annual Conference*, 1981, pp. 18–21.

15. Clark, D.R., Lehto, M.R., Reliability, maintenance, and safety of robots, *Handbook of Industrial Robotics*, ed. S.Y. Nof, John Wiley and Sons, New York, 1999, pp. 717–753.

16. *Safeguarding Industrial Robots, Part II: Welding and Allied Processes*, The Machine Tool Trades Association (MTTA), London, 1985.

7

Robot Accidents and Analysis

7.1 Introduction

An accident is an unplanned and undesired event; every year, thousands of lives are lost due to work-related injuries. For example, in 2007, in the United States, 5488 workers died from job-related injuries and about four million workers suffered from nonfatal work-related injuries or illnesses [1,2].

Needless to say, accidents are an important issue and, in regard to robots, there have been many fatal and nonfatal accidents over the years. For example, as per [3], up to the early 1980s, there were five fatal accidents involving robots: four in Japan and one in the United States; the up-to-date information on fatal and nonfatal robot accidents in the United States is available from [4].

This clearly indicates that robot-related accidents are pressing problems and thus require careful attention from industry. This chapter presents various important aspects of robot accidents and analysis.

7.2 Some Examples of Robot-Related Accidents

Over the years, many robot-related accidents have occurred. Some examples of such accidents are as follows [3,5–9]:

- A worker entered the robot cell to clean its sensors and was killed because he overlooked stated lockout procedures.
- A repair person climbed over a safety fence around the robot work area without turning off power to the robot and performed tasks in the area while the robot was temporarily stopped. When the robot recommenced movement, it pushed the repair person into a nearby grinding equipment and, consequently, the person died.
- A robot operator went to check or fix a robot failure without locking the robot. During the repair process, the operator activated the robot

and the arm of the robot crushed the operator against a part being transported on a conveyor and the operator died.

- A worker stepped between a machine (a planer) and the robot it was servicing, and turned off the circuit that was emitting activating signals from the planer to the robot. After carrying out the required task, the worker switched on the same circuit while still being in the workspace area of the robot. The robot started its operation and crushed the worker to death against the planer.
- A worker was freeing a jam at the end of an oven and became caught between the conveyor belt of the oven and the robotic arm and got killed.
- A worker turned on a welding robot while another worker was still in the robot's work area. The robot pushed the worker in its work area into the positioning fixture; the worker was consequently killed.
- A worker entered the caged robotic palletizer cell while the robotic palletizer was still running. The worker's torso was crushed by the arms of the robotic palletizer as it tried to pick up boxes on the roller conveyor and the worker died.
- A worker climbed onto a conveyor belt in motion for recovering a faulty part when the robot serving the line was stopped on a program point temporarily. The robot crushed the worker to death when the operation was restarted.
- A worker was troubleshooting a robotic arm used for removing CD jewel cases from an injection molding machine, when the arm cycled and struck the worker, who died two weeks later.
- A worker violated safety devices for entering a work cell while the material handling robot was operating it in automatic mode. The worker was trapped between the robot and a post anchored to the floor. The injured worker died a few days later.
- A worker freeing a jam at the end of an oven was caught between a robotic arm and the conveyor belt of the oven and was killed.
- A worker holding the robot controls in his hand activated the robot while bending over the wheel to check the settings. The robot pinned the worker against the wheel and crushed him to death.
- A repair person observing the operation of a robot entered the work area of another robot without being aware of it. When the other robot moved back to its home location, it knocked down the repair person who suffered a cervical strain as the result of being knocked down.
- During the manual operation, a robot's arm went out of sequence; consequently, when the robot operator tried to regulate it, the operator received a cut on the head.

- A worker assisting an electrical engineer in troubleshooting a malfunctioning robot was crushed between the lifting arm and the lower frame of the robot when the lifting arm dropped.

7.3 Robot Accidents: Causes and Sources

Over the years, many studies have been conducted to determine the causes for the occurrence of robot-related accidents. The findings of two studies are presented below.

7.3.1 Study I

A Japanese study based on a survey of robots reported the following causes (with corresponding percentages in parentheses) for 18 near-accidents [10]:

- Erroneous robot movement during testing or teaching (16.6%)
- Wrong action by the robot during manual operation (16.6%)
- Wrong movement of peripheral equipment during normal operation (5.6%)
- Wrong movement during repair, regulation, and checking (16.6%)
- Erroneous robot movement during normal operation (5.6%)
- Sudden entry of the human into the robot work area (11.2%)
- Wrong movement of peripheral equipment during testing or teaching (16.6%)
- Others (11.2%)

The above percentages, given in parentheses, indicate that robot accidents are most likely during setting, adjusting, or other manual operations. Some of the possible reasons for the occurrence of these accidents are as follows [5]:

- Workers frequently forget about hazards associated with robot systems under normal or abnormal conditions.
- Workers involved with robot systems become preoccupied and careless/complacent.
- Workers involved with robot systems often take chances as opposed to following the stated procedures carefully/vigilantly.

7.3.2 Study II

A General Motors Corporation study reported the following causes for many robot incidents [5]:

- The presence of an ignorant individual in the operating enclosure of the robot—people with authorization ignorant of the ramifications of the robot program
- Lack of vigilance from involved workers to adjacent robots

There are many possible sources for robot-related accidents, which may be divided into two groups as follows [5,11,12]:

- *Group I: Engineering factors.* This group includes factors such as failure of the control panel or peripheral equipment, poor software design, high speed of the robotic arm, the robot worker work environment, problems associated with the design of the gripper and/or the control panel, robot system failures (electrical, mechanical, etc.), inadvertent movement of the robot, unintentional contact with the start and other switches, and failure of safety devices.
- *Group II: Behavioral and organizational factors.* This group includes factors such as wrong procedures for initial start-up of the robot, programmed movement of the robot unknown to the robot operator, poor robot-related training programs for people such as repair persons, operators, programmers, and supervisory personnel, lack of awareness among operators of the work function of the robot or an operator being unaware that another person is within the danger space, and curiosity leading a nonoperating person to approach a robot.

It is to be noted that some of the above engineering, behavioral, and organizational factors cannot be strictly divided into one group or another. Furthermore, there could be, direct or indirect, overlap between various causes and sources for robot-related accidents. The main objective here is to highlight the direction that a safety professional should follow in further analyzing the underlying sources and causes of robot-related accidents.

7.4 Effects of Robot-Related Accidents

The effects of robot-related accidents on humans may range from no injury to a fatality. A study of 32 robot-related accidents grouped the effects of the accidents in the following two categories [13]:

- *Category I*: Pinch-point (i.e., a person or part of his/her body pinched between two robot parts or between the robot itself and some external item)
- *Category II*: Impact

The percentage breakdowns between these categories (i.e., pinch-point and impact) were roughly 56% and 44%, respectively. Some examples of the effects of robot-related accidents are as follows [5]:

- Pinch-point injury to person's chest
- Pinch-point injury to person's hand
- The arm of the robot strikes a person's head
- The arm of the robot programmer impacted by the robotic arm
- The arm of the robot strikes a person

The following workers were involved in the 32 robot accidents [5,13,14]:

- Line operators (23 times)
- Maintenance personnel (six times)
- Programmers (three times)

The breakdowns between impact and pinch among these workers were as follows [5,13]:

- *Impact*: Two (programmers), two (maintenance personnel), 10 (line operators)
- *Pinch*: One (programmers), four (maintenance personnel), 13 (line operators)

Finally, it is to be noted from the above data that the line operators were involved around 72% of the robot-related accidents.

7.5 Robot-Related Accidents at Manufacturer and User Facilities

Robot-related accidents can occur either during the development process at the manufacturer's facilities or during the commissioning period, and/or in use at the premises of the user [5,15]. Usually, accidents at the manufacturing facilities occur during the test run period or during the programming period. The main reason for the occurrence of accidents during the robot programming process is that people frequently stand within the work area of the robot.

The opportunities for the occurrence of a robot-related accident are quite high during the test run process, since there is no way to be absolutely sure that the program is totally bug-free or that the associated hardware is wholly reliable.

The likelihood of robot-related accidents at the facilities of the user is higher than at the manufacturer's facilities basically due to the following two reasons:

- The length of the robotic operation is much higher.
- The surroundings in which the robots carry out their assigned mission are subject to intrusion from people who may not possess proper knowledge of the safety needs of robots.

At the users' facilities, robot-related accidents may occur during programming, maintenance, or when robots carry out operations in the automatic mode. Normally, to mend robots, the programmers enter the work area of robots and this creates hazardous situations. During maintenance of a robot, interactions between the maintenance personnel and the robot are quite close, and any carelessness in safety precautions can result in an accident.

During the operations of robots in the automatic mode an accident can occur more as a result of human error than anything else, because intrusions inside work zones of robots in the automatic mode can be very risky. Additional information, directly or indirectly, concerned with robot-related accidents at user facilities is available in the preceding sections of this chapter.

7.6 Useful Recommendations to Prevent Human Injury by Robots

Over the years, many organizations around the world have proposed useful recommendations to prevent injury to humans by robots. The recommendations proposed by the National Institute for Occupational Safety and Health in the United States were directed at the following three areas [16]:

- Robot system design
- Supervision of workers
- Training of workers

The recommendations concerning each of the above three areas are presented below, separately.

7.6.1 Robot System Design

The following recommendations are concerned with both new designs and existing robotic equipment:

- Ensure that control and operational areas of the robot have appropriate levels of illumination.
- Ensure that adequate clearance distances are provided around all moving parts of the robot system.
- Ensure that appropriate remote "diagnostic" instrumentation is incorporated so that troubleshooting of the system can be performed mainly from places outside the work area of the robot.
- Ensure that proper physical barriers incorporating gates with electrical interlocks are provided so that, at the moment of opening the gate, the operation of the robot is stopped instantly.
- Ensure that floors or working surfaces contain adequate visible marks that show the work area of the robot clearly.
- Ensure that proper barriers between freestanding objects and the robot equipment are provided appropriately.
- Ensure that, as a back-up to motion sensors, floor sensors, light curtains, or electrical interlocks (that turns off the robot instantly in situations whenever the barrier is crossed by a person) are included.

7.6.2 Supervision of Workers

The following two recommendations are for the work supervisors:

- Recognize that, with time, experienced personnel carrying out automated tasks may become overconfident, complacent, or inattentive to the danger present in automated items.
- Ensure that no human enters the work area of the robot prior to putting the robot on "hold," or on a reduced speed mode.

7.6.3 Training of Workers

The following recommendations are concerned with training of workers:

- Ensure that training clearly emphasizes that workers involved with robots must be knowledgeable about all working aspects of the robot, prior to carrying out their assigned tasks at robot workstations.
- Provide appropriate training, specific to the particular robot under consideration, to operation personnel, programmers, and maintenance personnel.

- Ensure that training clearly emphasizes that all robot operators must never be within reach of operating robots.
- Ensure that training clearly emphasizes the operating of robots at reduced speeds—consistent with a satisfactory human response time to avoid risks during the programming process—by programmers, operators, and maintenance personnel and that they should be alert to each and every possible pinch-point within the robots' operational areas. Two typical examples of pinch-points are walls and poles.
- Provide appropriate refresher training courses to experienced operators, maintenance personnel, and programmers that emphasize new technological developments and robot-related safety.

7.7 Methods for Performing Robot Accident Analysis

There are many methods for analyzing robot-related accidents. Some of these methods are presented below.

7.7.1 Estimating Probability of an Accident Occurrence Related to the Operation of a Robot

This section presents an expression that can be used to estimate the probability of an accident occurring related to the operation of a robot. The probability of such an occurrence is expressed by [5]

$$P_{ar} = P_A + P_B + P_C \tag{7.1}$$

where
P_{ar} is the probability of an accident occurrence from the operation of a robot.

$$P_A = \sum \{P_1 + P_2(1 - P_{11})\}\{1 + P_3 + P_4\} \tag{7.2}$$

$$P_B = \sum \{P_5 + P_6(1 - P_{11})\}\{1 + P_7 + P_8\} \tag{7.3}$$

$$P_c = \sum (P_9 + P_{10})(1 - P_{11}) \tag{7.4}$$

The symbols used in the above three equations (i.e., Equations 7.2 through 7.4) are defined below:

P_1 is the probability of the occurrence of an irreversible human failure that could result in or allow an accident.

P_2 is the probability of the occurrence of a reversible human failure that could result in or permit a mishap.

P_3 is the probability of the robot possessing an adverse characteristic that could result in human failure (error).

P_4 is the probability of the robot experiencing an adverse environmental condition that could result in human failure (error).

P_5 is the probability of occurrence of the failures that could result in mishaps (in this case no corrective measure is possible).

P_6 is the probability of occurrence of those malfunctions that could result in accidents unless the appropriate actions are taken in a timely manner.

P_7 is the probability of the robot having an adverse characteristic that could result in human failure (error).

P_8 is the probability of the robot experiencing an adverse environmental condition that could result in robot failure.

P_9 is the probability of the robot possessing an adverse characteristic that could result in injury, damage, or loss in the absence of material failure or error.

P_{10} is the probability of the robot experiencing an adverse environmental condition that could lead to damage or injury in the absence of error or failure.

P_{11} is the probability of necessary action taken as required.

7.7.2 Root Cause Analysis

Root cause analysis (RCA) may simply be described as a systematic investigation approach that uses information collected during an assessment of an accident for determining the underlying factors for the deficiencies that resulted in the accident [17]. The method was originally developed by the U.S. Department of Energy to investigate industrial accidents [18,19].

The performance of the RCA helps in better understanding the causal factors in the sequence of evolving events; the RCA process is concluded with recommendations for improvements on the basis of the investigational findings [19,20].

The steps generally involved to perform RCA are as follows [20,21]:

- Educate everyone involved in RCA
- Inform appropriate staff members when the occurrence of a sentinel event is reported
- Form an RCA team by including appropriate individuals
- Prepare for and hold the first team meeting
- Determine the event sequence
- Separate and identify each event sequence that may have been a contributory factor in the sentinel event occurrence
- Brainstorm about the factors surrounding the selected events that may have, directly or indirectly, contributed to the occurrence of the sentinel event
- Affinitize with the brainstorming session results
- Develop the appropriate action plan
- Distribute the RCA document and the associated action plan to all individuals concerned

7.7.2.1 RCA Software

Over the years, various types of software packages have been developed for conducting RCA. The benefits cited for the automation of RCA include better data organization, enhanced follow-up capabilities, reduction in analysis time, improved rigor, and easier reporting capabilities [17]. Some of the available RCA software packages in the market are as follows [17]:

- BRAVO, developed by JBF Associates, Inc., 1000 Technology Drive, Knoxville, TN 37939, USA.
- REASON, developed by Decision Systems, Inc., 802 N. High St., Suite C, Longview, TX 75601, USA.
- PROACT, developed by Reliability Center, Inc., P.O. Box 1421, Hopewell, VA 23860, USA.
- TAPROOT, developed by System Improvements Inc., 238 S. Peters Road, Suite 301, Knoxville, TN 37923-5224, USA.

7.7.2.2 RCA Advantages and Disadvantages

Today, RCA is often used to investigate various types of industrial accidents. Some of its advantages are as follows [22]:

- It is an effective tool to address and identify organizational and systems issues.

- It is a well-structured and process-focused method.
- The systematic application of this method can uncover common root causes that link a disparate collection of accidents.

In contrast, some of the disadvantages of RCA are as follows [22]:

- It is quite a time-consuming and labor-intensive method.
- It is quite possible to be tainted by the bias of hindsight.
- It is impossible to determine precisely if the root cause established by the analysis is really the actual cause for the occurrence of the accident.
- In essence, RCA is an uncontrolled case study.

7.7.3 Fault Tree Analysis

The fault tree method was developed in the early 1960s to analyze the Minuteman Launch Control System with respect to safety at the Bell Telephone Laboratories [23]. Today, it is widely used throughout the world to perform various types of studies on reliability and safety. The method is described in Chapter 4.

Here, the application of the fault tree method to perform robot accident analysis is demonstrated through the following two examples:

EXAMPLE 7.1

Assume that at a robotic facility, a robot-related accident involving a human, caused by sudden robot movement, can occur due to the occurrence of the following: human in the robot's work area and sudden robot movement. The event "human in robot work area" can be either intentional entry or unintentional entry. The event "sudden robot movement" can occur due to the occurrence of the following: power supply on and unexpected start signal.

In turn, the event "power supply on" can occur either due to control mechanism malfunction or human error. With the aid of the fault tree symbols given in Chapter 4, develop a fault tree for the top event "robot-related accident, involving a human, caused by sudden robot movement."

A fault tree for the example is shown in Figure 7.1. The single capital letters in the figure denote corresponding fault events (e.g., A: unexpected start signal, Y: Human in robot work area, and Z: Power supply on).

EXAMPLE 7.2

Assume that in Figure 7.1, the probabilities of occurrence of fault events A, B, C, D, and E are 0.05, 0.1, 0.04, 0.15, and 0.06, respectively. With the aid of Chapter 4, calculate the probability of occurrence of the top

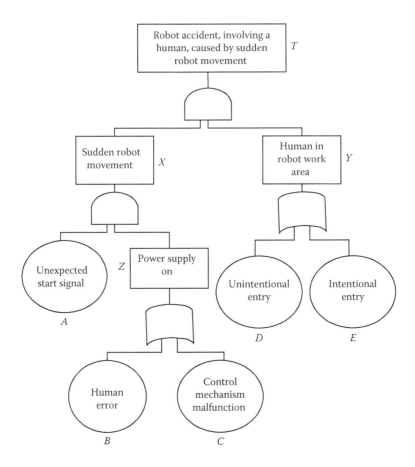

FIGURE 7.1
A fault tree for the top event: Robot-related accident, involving a human, caused by sudden robot movement.

event *T*: Robot-related accident, involving a human, caused by sudden robot movement. With the aid of material presented in Chapter 4, we calculate the probabilities of occurrence of fault events *Z, Y, X*, and *T* as follows:

The probability of the occurrence of event *Z* is

$$P(Z) = 1 - (P(B))(1 - P(C))$$

$$= 1 - (1 - 0.1)(1 - 0.04)$$

$$= 0.136 \tag{7.5}$$

where
P(B) is the occurrence probability of event *B*.
P(C) is the occurrence probability of event *C*.

Similarly, the probability of the occurrence of the event Y is

$$P(Y) = 1 - (1 - P(D))(1 - P(E))$$

$$= 1 - (1 - 0.15)(1 - 0.06)$$

$$= 0.201 \tag{7.6}$$

where
$P(D)$ is the occurrence probability of event D.
$P(E)$ is the occurrence probability of event E.

The probability of the occurrence of the intermediate event X is

$$P(X) = P(A)P(Z)$$

$$= (0.05)(0.136)$$

$$= 0.0068 \tag{7.7}$$

where
$P(A)$ is the occurrence probability of event A.

The top event T (robot-related accident, involving a human, caused by sudden robot movement) probability of occurrence is

$$P(T) = P(X)P(Y)$$

$$= (0.0068)(0.201)$$

$$\cong 0.0014 \tag{7.8}$$

Thus, the probability of the occurrence of the top event T (robot-related accident, involving a human, caused by sudden robot movement) is approximately 0.0014. The Figure 7.1 fault tree with calculated and given probability values is shown in Figure 7.2.

7.7.4 Markov Method

The Markov method is widely used to perform various types of reliability and availability analyses and is described in Chapter 4. Here, the application of the Markov method to perform robot-related accident analysis is demonstrated through the following two examples:

EXAMPLE 7.3

Assume that a robot system can fail safely or with an accident. The robotic system's safe failure rate is λ_s and its failing with an accident failure rate is λ_a. The robot system state-space diagram is shown in Figure 7.3. The numerals in the diagram circle and boxes denote the robot system states. Develop expressions for the robot system state probabilities and mean

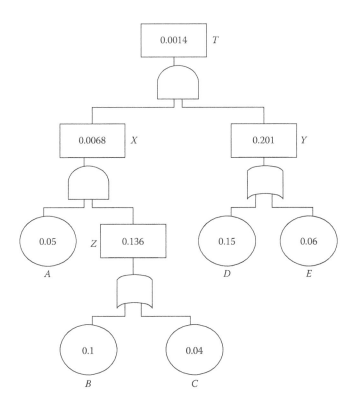

FIGURE 7.2
Redrawn Figure 7.1 fault tree with calculated and given event occurrence probability values.

time to failure by using the Markov method and assuming that robot system failures occur independently and its failure rates are constant.
    The following symbols are associated with the Figure 7.3 state-space
diagram and its associated equations:

    j is the state of the robot system: $j = 0$ means the robot system is
        operating normally; $j = 1$ means the robot system has failed
        safely; $j = 2$ means the robot system has failed with an accident.
    t is time.

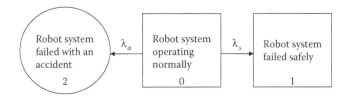

FIGURE 7.3
Robot system state-space diagram.

$P_j(t)$ is the probability that the robotic system is in state j at time t, for $j = 0,1,2$.

λ_s is the constant robot system, failing safely, failure rate.

λ_a is the constant robot system, failing with an accident, failure rate.

By using the Markov method presented in Chapter 4, we write down the following set of differential equations for the Figure 7.3 diagram:

$$\frac{dP_0(t)}{dt} + P_0(t)(\lambda_s + \lambda_a) = 0 \tag{7.9}$$

$$\frac{dP_1(t)}{dt} - P_0(t)\lambda_s = 0 \tag{7.10}$$

$$\frac{dP_2(t)}{dt} - P_0(t)\lambda_a = 0 \tag{7.11}$$

At time $t = 0$, $P_0(0) = 1$, $P_1(0) = 0$, and $P_2(0) = 0$.
By solving Equations 7.9 through 7.11, we get

$$P_0(t) = e^{-(\lambda_s + \lambda_a)t} \tag{7.12}$$

$$P_1(t) = \frac{\lambda_s}{\lambda_s + \lambda_a}\left[1 - e^{(\lambda_s + \lambda_a)t}\right] \tag{7.13}$$

$$P_2(t) = \frac{\lambda_a}{\lambda_s + \lambda_a}\left[1 - e^{-(\lambda_s + \lambda_a)t}\right] \tag{7.14}$$

By integrating Equation 7.12 over the interval $[0,\infty]$, we get the following expression for the robot system mean time to failure [24]:

$$MTTF_{rs} = \int_0^\infty e^{-(\lambda_s + \lambda_a)t}\,dt$$

$$= \frac{1}{\lambda_s + \lambda_a} \tag{7.15}$$

where
$MTTF_{rs}$ is the robot system mean time to failure.

EXAMPLE 7.4

Assume that the robotic system's failing with an accident failure rate is 0.0002 failures per hour and its failing safely failure rate is 0.0005 failures per hour. Calculate the robot system failure rate failing with an accident probability during a 500-h mission and its mean time to failure.

By substituting the specified data values into Equation 7.14, we obtain

$$P_2(500) = \frac{0.0002}{0.0005 + 0.0002}\left[1 - e^{-(0.0005+0.0002)(500)}\right]$$
$$= 0.0843$$

Similarly, by inserting the specified data values into Equation 7.15, we get

$$MTTF_{rs} = \frac{1}{0.0005 + 0.0002}$$
$$= 1428.57 \text{ h}$$

Thus, the robotic system failing with an accident probability and mean time to failure are 0.0843 and 1428.57 h, respectively.

7.8 Problems

1. Discuss the causes for the occurrence of robot-related accidents.
2. What are the sources of robot-related accidents?
3. Discuss the effects of robot-related accidents with regard to humans.
4. Discuss robot-related accidents at the robot manufacturer's facilities.
5. Discuss robot-related accidents at the robot user's facilities.
6. Discuss recommendations to prevent human injury by robots.
7. What are the general steps involved to perform root cause analysis?
8. What are the advantages and disadvantages of root cause analysis?
9. Prove Equations 7.12 through 7.14 with the aid of Equations 7.9 through 7.11.
10. Assume that a robot system can fail safely or with an accident and its failing with an accident failure rate is 0.0004 failures per hour and its failure rate in failing safely is 0.0009 failures per hour. Calculate the probability for the robot system failing safely during a 200-h mission and its mean time to failure.

References

1. Steenland, K., Burnett, C., Lalich, N., Ward, E., Hurrall, J., Dying for work: The magnitude of U.S. mortality from selected causes of death associated with occupation, *American Journal of Industrial Medicine*, Vol. 43, 2003, pp. 461–482.

2. Bureau of Labor Statistics, United States Department of Labor, *Workplace Injuries and Illnesses in 2007*, Washington, DC, 2008.

3. Altamuro, V.M., *Working Safely with the Iron Collar Worker*, National Safety News, July 1983, pp. 38–40.

4. United States Department of Labor, Occupational Safety and Health Administration (OSHA), *Reports No. 0626300, 0950636, 0728900, 0627100, 0111700, 0454712, and 0552652*, 200 Constitution Avenue, Washington, DC.

5. Dhillon, B.S., *Robot Reliability and Safety*, Springer-Verlag, New York, 1991.

6. United States Department of Labor, Occupational Safety and Health Administration (OSHA), *Report No. 0552652*, Washington, DC, October 10, 2006.

7. Nicolaisen, P., Safety problems related to robots, *Robotics*, Vol. 3, 1987, pp. 205–211.

8. Lauch, K.E., New standards for industrial robot safety, *CIM Review*, Spring 1986, pp. 60–68.

9. Japanese Ministry of Labor, Tokyo, *Study on Accidents Involving Industrial Robots, Report No. PB83239822*, 1982; available from the National Technical Information Service (NTIS), Springfield, Virginia 22161.

10. Sugimoto, N., Safety engineering on industrial robots and their draft standard safety requirements, *Proceedings of the 7th International Symposium on Industrial Robots*, October 1977, pp. 461–468.

11. Rahimi, M., System safety approach to robot safety, *Proceedings of the Human Factors Society's 28th Annual Meeting*, 1984, pp. 102–106.

12. Rahimi, M., Systems safety for robots: Energy barrier analysis, *Journal of Occupational Accidents*, Vol. 8, 1986, pp. 127–138.

13. Jiang, B.C., Gainer, C.A., A cause and effect analysis of robot accidents, *Journal of Occupational Accidents*, Vol. 9, 1987, pp. 27–45.

14. Hirschfeld, R.A., Aghazadeh, F., Chapleski, R.C., Survey of robot safety in industry, *The International Journal of Human Factors in Manufacturing*, Vol. 3, No. 4, 1993, pp. 369–379.

15. Ramachandran, V., Vajpayee, S., Safety in robotic installations, *Robotics and Computer-Integrated Manufacturing*, Vol. 3, 1987, pp. 301–309.

16. National Institute for Occupational Safety and Health, *Request for Assistance in Preventing the Injury of Workers by Robots, Report No. PB85-236818*, Cincinnati, Ohio, December 1984; available from the National Technical Information Service (NTIS), Springfield, Virginia.

17. Latino, R.J., Automating root cause analysis, in *Error Reduction in Health Care*, ed. P.L. Spath, John Wiley and Sons New York, 2000, pp. 155–164.

18. Lement, B.S., Ferrera, J.J., Accident causation analysis by technical experts, *Journal of Product Liability*, Vol. 5, No. 2, 1982, pp. 145–160.

19. Dhillon, B.S., *Safety and Human Error in Engineering Systems*, CRC Press, Boca Raton, Florida, 2013.

20. Dhillon, B.S., *Human Reliability and Error in Medical System*, World Scientific Publishing, River Edge, New Jersey, 2003.

21. Burke, A., *Root Cause Analysis, Report*, 2000; available from the Wild Iris Medical Education, P.O. Box 257, Comptche, California.

22. Wald, H., Shojania, K.G., Root causes analysis, in *Making Healthcare Safer: A Critical Analysis of Patient Safety Practices*, ed. A.J. Markowitz, Report No. 43,

Agency for Healthcare Research and Quality, U.S. Department of Health and Human Services, Rockville, MD, 2001, Chapter 5, pp. 1–7.

23. Dhillon, B.S., Singh, C., *Engineering Reliability: New Techniques and Applications*, John Wiley and Sons, New York, 1981.
24. Dhillon, B.S., *Design Reliability: Fundamentals and Applications*, CRC Press, Boca Raton, Florida, 1999.

8

Robot Maintenance and Areas of Robotics Applications in Maintenance and Repair

8.1 Introduction

Robots are quite complex machines that can sense information from their surrounding environment, autonomously make effective decisions, and manipulate the surrounding environment with effectors. Just as in the case of any other complex engineering systems, robots also require preventive and corrective maintenance. Therefore, it is very important that users of robots devise an effective maintenance program as, otherwise, the unscheduled downtime of robots may increase. Furthermore, a poor maintenance program may result in safety problems as well.

Needless to say, both robot users and manufacturers play an important role in the maintenance aspects of robots as careful consideration to maintenance is given not only during the operational phase of robots but during the design phase too.

Today, robots are being used in a variety of applications in maintenance and repair quite successfully, among these the nuclear industry, railways, in maintenance of power lines, highways, and in aircraft servicing. Needless to say, this chapter presents various different aspects of robot maintenance and maintenance and repair in different areas of robotics applications.

8.2 Robot Maintenance-Related Needs and Maintenance Types

Generally, the maintenance needs of a robot are determined by its specific application and type. The robot powering system is probably the most important factor that clearly affects its maintenance. Most of the robots used in industry sectors may be classified under the following two categories [1]:

- *Hydraulic with electrical controls.* These are robots whose working parts are driven hydraulically and, generally, are controlled by electrical components or parts.

- *Electrical.* These are robots that are driven and controlled by electrically powered parts.

Nonetheless, it is to be noted that irrespective of the robot type, the mechanical parts of a robot require proper attention. Normally, as per [2], the following maintenance-related tasks are relevant to robots:

- Attending to seals and replacing protective accessories
- Inspecting all involved parts regularly for wear, particularly wrists
- Cleaning to eradicate corrosive agents
- Lubricating

Specifically, in the case of hydraulic robots, the items that need proper attention include bearings, servo valves, filters, hydraulic oil, and high-pressure hoses.

Three basic types of maintenance for robots used in industrial applications are corrective maintenance, preventive maintenance, and predictive maintenance. Corrective maintenance is concerned with repairing the robot system to return its operational state whenever it breaks down. Preventive maintenance is basically concerned with servicing robot system parts regularly. Finally, predictive maintenance is concerned with predicting when a failure might occur and to alert personnel involved with maintenance. Predictive maintenance is becoming increasingly effective as modern robot systems are equipped with highly sophisticated electronic parts and sensors.

8.3 Commonly Used Tools to Maintain a Robot and Measuring Instruments and Tooling for Periodic Robot Inspections

Various types of tools are used to maintain a robot system. They range from diagnostic codes displayed on the robot control panel to wrenches and specific maintenance manuals. Although the type of maintenance tools required will be peculiar to the specific robot system involved, some of the commonly used tools to maintain a robot system are as follows [1]:

- Torque wrenches
- Circuit-card pullers
- Alignment fixtures

- Seal compressors
- Accumulator charging adaptors

Usually, a variety of measuring instruments and tooling are required to perform periodic inspections on robots. Some examples of measuring instruments are slide calipers, the direct current (DC) voltmeter, the alternating current (AC) voltmeter, and the oscilloscope [3]. Tooling includes items such as the following [3]:

- Large, medium, and small cross-headed screwdrivers
- Grease gun
- Medium and small monkey wrenches
- Double-head wrench
- Large, medium, and small screwdrivers
- Nippers
- Radio pliers
- Pincers
- C-shaped snap-ring pliers
- Hexagon wrench kit

8.4 Robot Diagnosis and Monitoring Approaches

In the industrial sectors, many different methods are used for monitoring the robotics equipment status, diagnosing their conditions, and then initiating appropriate maintenance-related activities, such as overhauls, repair, and replacement, when necessary [4]. In this regard, condition-based maintenance policies typically schedule inspections by the involved maintenance or operating personnel on a regular basis. Control charts and other approaches used in the area of quality control are also utilized for determining the degree of deterioration over a period of time in production and for consequent equipment maintenance activities such as replacing parts, recalibration, nonroutine inspections, and adjustments.

Other commonly used diagnostic approaches are vibration analysis, behavior monitoring as well as monitoring of structural integrity, wear debris (i.e., contaminants) hydraulic fluids and lubricants, and so on. Finally, it is to be noted that industrial robots generally contain internal diagnostic functions, which are considered quite useful during troubleshooting and highlight the need to perform maintenance on the machine/system [4].

8.5 Useful Guidelines to Safeguard Robot Maintenance Personnel and Safeguarding Methods for Use during the Robot Maintenance Process

Although there are many guidelines to safeguard robot maintenance personnel, the guidelines considered most useful are as follows [5]:

- Ensure that the entire robot system is turned off during the maintenance and repair process, and all the power sources and the releasing of potentially dangerous stored energy are properly locked out/tagged.
- Ensure that all involved maintenance personnel are properly trained in regard to procedures essential to perform all the required maintenance-related tasks safely and effectively.
- Ensure that in the event of a lockout/tagout procedure, equally effective alternative safeguarding methods are employed.
- Ensure that all involved maintenance personnel are properly safeguarded from unintended or unexpected motion.

In situations when power cannot be turned off during the robot system maintenance process, the safeguarding methods shown in Figure 8.1 can be used [6].

The robot arm predetermined position method is concerned with placing the arm of the robot in a predetermined position so that necessary maintenance can be performed without exposing involved personnel to trapping points. The emergency stop method is concerned with making the emergency stop readily accessible, as well as making the restarting of the robot impossible until the time that the emergency stop device is reset manually.

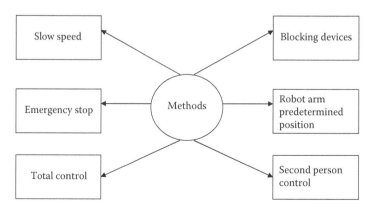

FIGURE 8.1
Safeguarding methods for use during the robot system maintenance process.

The total control method is basically concerned with placing the complete control of the robot system in the hands of maintenance personnel. Thus, in this method, the requirements such as those presented below must be considered:

- All robot system-associated emergency stop devices are functional.
- All remote interlocks that can cause robot movement in automatic mode are subject to control override.
- In situations where the other equipment movement in a robot system is hazardous to humans in the restricted work enclosure, the control of such movement is to be placed in the hands of all those individuals.
- The robot system is reset in automatic mode only after the departure of all individuals from the work enclosure.

The slow-speed method is concerned with reducing the speed of robot to a slow-speed level for maintenance operations. The second person control method calls for the presence of a second person (i.e., a person who is knowledgeable about potential hazards associated with a robot system and is also capable of reacting fast to protect others in the moment of need) at the robot system controls.

Finally, the blocking devices method is concerned with using devices such as blocks and pins during the maintenance process for thwarting potentially hazardous moments of the robot system.

8.6 Models for Performing Robot Maintenance Analysis

This section presents four mathematical models considered useful, directly or indirectly, to perform robot maintenance analysis.

8.6.1 Model I

This mathematical model is concerned with estimating the optimum number of inspections per robot facility per unit time. This type of information is important to decision makers because inspections are often disruptive; however, such inspections generally help to reduce the robot downtime because they result in fewer breakdowns. In this model, the total downtime of the robot is minimized to obtain the optimum number of inspections.

The total robot downtime per unit time is expressed by [7,8]

$$T_{rd} = kDT_i + \frac{(cDT_b)}{k} \tag{8.1}$$

where

T_{rd} is the total robot downtime per unit time.
DT_i is the downtime per inspection for a robot facility.
c is a constant for a specific robot facility.
k is the number of inspections per robot facility per unit time.
DT_b is the downtime per breakdown for a robot facility.

After differentiating Equation 8.1 with respect to k and then equating it to zero, we obtain

$$k^* = \left[\frac{(cDT_b)}{DT_i} \right]^{1/2}$$

(8.2)

where

k^* is the optimum number of inspections per robot facility per unit time.

By inserting Equation 8.2 into Equation 8.1, we obtain

$$T_{rd}^* = 2[cDT_bDT_i]^{1/2}$$

(8.3)

where

T_{rd}^* is the minimum total robot downtime.

EXAMPLE 8.1

Assume that for a certain robot facility, the following data values are given:

$$c = 2$$
$$DT_b = 0.10 \text{ month}$$
$$DT_i = 0.02 \text{ month}$$

Using the above-given data values, calculate the optimum number of monthly robot inspections and the minimum total robot downtime.

By inserting the above-given data values into Equations 8.2 and 8.3, we obtain

$$k^* = \left[\frac{\{2(0.10)\}}{(0.02)} \right]^{1/2}$$
$$= 3.16 \text{ inspections per month}$$

and

$$T_{rd}^* = 2[2(0.10)(0.02)]^{1/2}$$
$$= 0.126 \text{ month}$$

Thus, the optimum number of monthly robot inspections and the minimum total robot downtime are 3.16 and 0.126 month, respectively.

8.6.2 Model II

This mathematical model is concerned with maximizing the income of a robot subject to failure and repair. Using the Markov method given in Chapter 4 and [1,9], we write the following expressions for availability and unavailability of a robot:

$$AV_r(t) = \frac{\mu_r}{\lambda_r + \mu_r} + \frac{\lambda_r}{\lambda_r + \mu_r} e^{-(\lambda_r + \mu_r)t} \tag{8.4}$$

$$UAV_r(t) = \frac{\lambda_r}{\lambda_r + \mu_r}[1 - e^{-(\lambda_r + \mu_r)t}] \tag{8.5}$$

where
$UAV_r(t)$ is the robot unavailability at time t.
$AV_r(t)$ is the robot availability at time t.
λ_r is the robot constant failure rate.
μ_r is the robot constant repair rate.

For very large time t, Equations 8.4 and 8.5 reduce to the following equations:

$$AV_r = \lim_{t \to \infty} AV_r(t) = \frac{\mu_r}{\lambda_r + \mu_r} \tag{8.6}$$

$$UAV_r = \lim_{t \to \infty} UAV_r(t) = \frac{\lambda_r}{\lambda_r + \mu_r} \tag{8.7}$$

where
AV_r is the robot steady-state availability.
UAV_r is the robot steady-state unavailability.

Since $\lambda_r = 1/MTTF_r$ and $\mu_r = 1/MTTR_r$, Equations 8.6 and 8.7 become

$$AV_r = \frac{MTTF_r}{MTTF_r + MTTR_r} \tag{8.8}$$

where
$MTTF_r$ is the robot mean time to failure.
$MTTR_r$ is the robot mean time to repair.

and

$$UAV_r = \frac{MTTR_r}{MTTF_r + MTTR_r} \tag{8.9}$$

More specifically, it is to be noted that Equation 8.8 yields the probability that the robot is functional (or a fraction of the time that the robot repair personnel are idle) and Equation 8.9 offers the probability that the robot is under repair (or a fraction of the time that the robot repair personnel are working).

Thus, the monthly cost of the robot maintenance personnel is expressed by

$$C_{rm} = \theta_r \mu_r = \frac{\theta_r}{MTTR_r} \tag{8.10}$$

where
θ_r is the robot maintenance cost (constant) dependent on the nature of the robot.

The expected monthly income from the robot output is expressed by

$$EMI_r = (MI_{rf})(AV)$$
$$= \frac{MI_{rf}(MTTF_r)}{MTTF_r + MTTR_r} \tag{8.11}$$

where
MI_{rf} is the monthly income from the robot output, if the robot worked full-time.

Thus, the robot's net income is given by

$$NI_r = EMI_r - C_{rm}$$
$$= \frac{MI_{rf}(MTTF_r)}{MTTF_r + MTTR_r} - \frac{\theta_r}{MTTR_r} \tag{8.12}$$

To maximize the robot net income, we differentiate Equation 8.12 with respect to $MTTR_r$, and then set the resulting derivatives equal to zero:

$$\frac{d(NI_r)}{d(MTTR_r)} = \frac{MI_{rf}(MTTF_r)}{(MTTF_r + MTTR_r)^2} + \frac{\theta_r}{(MTTR_r)^2} = 0 \tag{8.13}$$

By rearranging Equation 8.13, we get

$$MTTR^* = \frac{MTTF_r}{[(MI_{rf}(MTTF_r))/\theta_r]^{1/2} - 1} \tag{8.14}$$

where

$MTTR_r^*$ is the optimum mean time needed to repair the robot.

Inserting Equation 8.14 into Equations 8.8, 8.10, and 8.11, respectively, results in

$$AV_r^* = 1 - \left[\frac{\theta_r}{MI_{rf}(MTTF_r)} \right]^{1/2} \tag{8.15}$$

$$C_{rm}^* = \left[\frac{\theta_r MI_{rf}}{MTTF_r} \right]^{1/2} - \frac{\theta_r}{MTTF_r} \tag{8.16}$$

and

$$EMI_r^* = MI_{rf} - \left[\frac{\theta_r MI_{rf}}{MTTF_r} \right]^{1/2} \tag{8.17}$$

where

AV_r^*, C_{rm}^*, and EMI_r^* are the optimum values of AV_r, C_{rm}, and EMI_r, respectively.

8.6.3 Model III

This mathematical model is concerned with a robot that is subjected to a periodical inspection–repair process at a time interval of, say, T. The model assumes that at the start of the time interval T, the robot is in good condition, but during the interval it could malfunction. Thus, the probability of robotic failure during the time interval T is expressed by [1,10,11]

$$F_r(T) = \int_0^T f(x)dx \tag{8.18}$$

where

$f(x)$ is the probability density function associated with the time between robot malfunctions.

The mean or average availability of the robot over the time interval T is expressed by

$$AV_m = R_r(T) + \frac{X^*}{T} F_r(T) \tag{8.19}$$

where
$R_r(T) = 1 - F_r(T)$ is the probability that the robot will not malfunction during the time interval T.

X^* is the time at which the robot malfunction or failure occurs (i.e., within the time interval T).

The expected value of X^* is expressed by

$$E(X^*) = \frac{1}{F_r(T)} \int_0^T x\, f(x)dx$$

$$= \frac{TF(T)}{F_r(T)} - \frac{1}{F_r(T)} \int_0^T F(x)dx \tag{8.20}$$

where
$F(x)$ is the cumulative distribution function.

Taking the expected value of Equation 8.19, we obtain the following equation for the robot's expected mean or average availability:

$$E(AV_m) = R_r(T) + \frac{F_r(T)E(X^*)}{T} \tag{8.21}$$

By inserting Equation 8.20 into Equation 8.21, we get

$$E(AV_m) = [1 - F_r(T)] + \frac{F_r(T)}{T}\left[\frac{TF(T)}{F_r(T)} - \frac{1}{F_r(T)}\int_0^T F(x)dx\right]$$

$$= 1 - \frac{1}{T}\int_0^T F(x)dx \tag{8.22}$$

It is to be noted that Equation 8.22 is subject to the assumption that the duration of the inspection–repair process is negligible, which may or may not be realistic. Nonetheless, a more realistic assumption for a regular inspection–repair could be a duration of, say, time T_r. In such a situation, the net availability of the robot is expressed by

$$AV_{nr} = \frac{(AV_m)T}{T + T_r} \tag{8.23}$$

Thus, the expected value of Equation 8.23 is

$$E(AV_{nr}) = \left(\frac{T}{Y}\right)E(AV_m) \tag{8.24}$$

where
$Y = T + T_r$

By inserting Equation 8.22 into Equation 8.24, we get

$$E(AV_{nr}) = \left(\frac{T}{Y}\right)\left[1 - \frac{1}{T}\int_0^T F(x)dx\right]$$

$$= \frac{1}{Y}\left[T - \int_0^T F(x)dx\right] \tag{8.25}$$

EXAMPLE 8.2

Assume that the mean time to failure of a robot is 400 h and its failure times are exponentially distributed. After every 200 h of service, the robot is scheduled for periodic maintenance.

Calculate the robot's expected average availability over a time interval of 200 h under the following assumptions:

- Each and every inspection and repair restores the robot to a condition that is considered as good as new.
- The robot is repaired only at the scheduled time.
- The duration of time taken by the inspection–repair process is negligible.

Thus, in this case, the robot's failures are expressed by the following probability density function:

$$f(x) = \frac{1}{400}e^{-x/400}, \quad x \geq 0 \tag{8.26}$$

where
x is the time to failure random variable.

With the aid of the definition for cumulative distribution function given in Chapter 2, we obtain the following cumulative distribution function of the robot failure time:

$$F(x) = \frac{1}{400}\int_0^x e^{-x/400}dx$$

$$= 1 - e^{-x/400} \tag{8.27}$$

By inserting Equation 8.27 into Equation 8.22, we get

$$E(AV_m) = 1 - \frac{1}{200} \int\limits_{0}^{200} \left[1 - e^{-x/400}\right] dx$$

$$= 1 - \frac{1}{200}\left[x + 400e^{-x/400}\right]_{0}^{200}$$

$$= 0.7869$$

Thus, the expected average availability of the robot is 0.7869.

8.6.4 Model IV

This mathematical model is concerned with determining the robot's optimum inspection frequency to minimize the robot downtime per unit time. In this model, the total robot downtime (per unit time) is the function of inspection frequency and is expressed by [12,13]

$$RTDT(k) = RDT_i + RDT_r$$

$$= \frac{k}{\alpha} + \frac{\lambda_r(k)}{\mu_r} \tag{8.28}$$

where
RDT_i is the robot downtime due to per-unit-of-inspection.
RDT_r is the robot downtime due to per-unit-of-time repairs.
$RTDT(k)$ is the robot total downtime per unit of time.
$1/\alpha$ is the mean value of exponentially distributed inspection times.
$\lambda_r(k)$ is the robot failure rate.
μ_r is the robot repair rate.
k is the inspection frequency.

By differentiating Equation 8.28 with respect to k, we obtain

$$\frac{dRTDT(k)}{dk} = \frac{1}{\alpha} + \frac{1}{\mu_r} \cdot \frac{d\lambda_r(k)}{dk} \tag{8.29}$$

By setting Equation 8.29 equal to zero and then rearranging, we get

$$\frac{d\lambda_r(k)}{dk} = -\frac{\mu_r}{\alpha} \tag{8.30}$$

It is to be noted that the value of k will be optimum when the right and left sides of Equation 8.30 are equal. At this very point, the robot total downtime will be minimal.

EXAMPLE 8.3

Assume that the failure rate of a robot is expressed by

$$\lambda_r(k) = \lambda e^{-k} \tag{8.31}$$

where
λ is the robot failure rate at $k = 0$.

Obtain an expression for the optimal value of k with the aid of Equation 8.30 and then—by setting the value of $1/\mu_r = 0.04$ month, $1/\alpha = 0.01$ month, and $\lambda = 2$ failures per month—calculate the optimal value of the inspection frequency, k.

By inserting Equation 8.31 into Equation 8.30, we obtain

$$-\lambda e^{-k} = -\frac{\mu_r}{\alpha} \tag{8.32}$$

By rearranging Equation 8.32, we get

$$k* = \ln\left[\frac{\lambda\alpha}{\mu_r}\right] \tag{8.33}$$

where
$k*$ is the optimal value of inspection frequency, k.

By substituting the given data values into Equation 8.33, we get

$$k* = \ln\left[\frac{2(0.04)}{0.01}\right]$$
$$= 2.08 \text{ inspections per month}$$

Thus, the optimal value of the inspection frequency is 2.08 inspections per month.

8.7 Areas of Robotics Applications in Maintenance and Repair

Over the years, robots—whether teleoperated, under supervisory control, or autonomous—have been used in or are being considered for many different applications in the areas of maintenance and repair. Some of the areas of robotics applications in maintenance and repair are as follows [14]:

- Nuclear industry
- Highways

- Railways
- Underwater facilities
- Power line maintenance
- Aircraft servicing

The first five of the above robotics applications are described below and additional information on the remaining one is available in [14].

8.7.1 Nuclear Industry

Over decades, teleoperated robots have been well utilized in the area of maintenance in the nuclear industry. There are several features of maintenance that make their applications quite attractive in this arena. Some of these features are as follows:

- Repair and maintenance need high levels of dexterity.
- The low frequency of the operation clearly calls for a general-purpose system that is capable of performing an array of maintenance-related tasks.
- The complexity of the tasks could be quite unpredictable due to the uncertain impact of a failure.

Although humans are likely to be able to accomplish tasks faster than teleoperators, the use of teleoperators can be quite useful to

- Lower health-related risks and improve safety
- Improve availability by allowing the maintenance activity to take place during ongoing operations instead of halting ongoing operations
- Reduce mean time to repair by shortening the response time to failures

The main application areas of teleoperated robots in nuclear maintenance are as follows [14,15]:

- *Maintenance in nuclear reactors.* As nuclear reactors are less likely to be designed for robotic maintenance than industrial nuclear facilities, it often calls for innovative approaches to gain access to trouble areas. In the past, such factors have restricted the application of robots in this arena. However, owing to increasingly stringent limits on worker exposure to ionizing radiation and exposure reductions, increasing attention is being given to use robots for maintenance activities.

Additional information on this is available in [15–21].

- *Operation and maintenance of industrial nuclear facilities and laboratories.* Additional information on this area is available in [14–16,22].
- *Emergency intervention.* Additional information on this area is available in [23–25].
- *Dismantling and decommissioning nuclear facilities.* Additional information on this area is available in [26–30].

8.7.2 Highways

Highways have become important in the transportation network around the globe. Over the years, the increasing volume of traffic on the roadways causes deterioration of roadways due to poor maintenance. Robotics solutions in highway maintenance applications are becoming increasingly attractive because of their potential for lowering labor costs, improving the safety of the highway workers, increasing the repair quality, reducing delays in traffic flow, and so on.

Some of the applications in which robotics can be used in this specific area are shown in Figure 8.2 [14–31].

Although, over the years, relatively few implementations have been attempted in the area of robotics in highway maintenance and repair, many

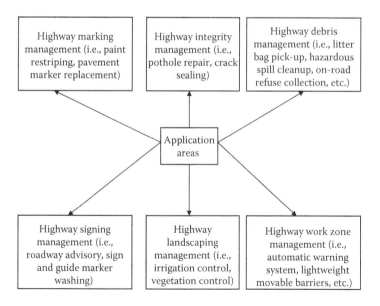

FIGURE 8.2
Robotic application areas in highway maintenance.

successful prototypes have been developed [32]. Some examples of robotics developments for the highway maintenance areas are as follows:

- Automated pavement distress data-collection vehicles [14,32]
- An automated cone dispenser that can effectively dispense and remove upto two rows of warning cones, for a total of about 240 cones [14,32]
- A rapid runway repair telerobitic system to repair craters on runways [14,33]
- A mobile lane separator that can place and remove concrete road marker blocks at speeds of upto 30 km/h [14]

8.7.3 Railways

Over the years, the railroad sector has recognized the economic advantages of automation, resulting in the development of a number of robotics solutions for the maintenance and repair in railways. The most common locations of robots are the railway maintenance workshops, which undertake activities such as cleaning, painting, and grinding [34]. Two examples of robotics developments in railway maintenance are as follows:

- The development of an automated system for cleaning the undersides of subway cars; the system involves the application of three industrial painting robots installed on either side and under the subway car being cleaned [35]. In this case, an operator starts the cleaning operations and then the complete system is controlled by a master computer that supervises the three individual robot controllers.
- The development of the RMS-2 rail grinding system that has automated capabilities for sensing the existing condition of the rails' surface [36]. The system can grind rails at speeds upto six miles per hour and contains 120 stones for grinding, spread along the underside of five of the 12 railway cars that make up the total system.

8.7.4 Underwater Facilities

In the offshore oil excavation sector, teleoperated robots are often used for maintaining facilities beneath the ocean surface. Their specific applications include repairing items such as pipelines, platforms, communications cables, and well heads [14].

Teleoperated robots are also used to inspect and repair water conveyance tunnels, inspect and clean steam generators, carry out underwater construction, clean marine growth from power plant cooling systems, and so on

[37–39]. As per [40], there are around 180 different commercially available, remotely controlled systems and around 63 companies are involved in building remotely operated manipulators or vehicles for subsea work. Needless to say, there is a wide range of remotely operated systems and additional information on the topic is available in [14,37–39].

8.7.5 Power Line Maintenance

Human operators perform maintenance-related tasks on power lines such as replacing ceramic insulators that support conductor wire as well as opening and enclosing the circuit between poles on overhead transmission lines. In order to replace humans in performing tasks such as these, many electric power companies have, since the 1980s, been investigating the use of robotic systems for live-line power line maintenance. Some examples of the robotics developments in power line maintenance are as follows:

- Teleoperator for operations, maintenance, and construction using advanced technology (TOMCAT) [41]
- A two-manipulator telerobotic system for live-line maintenance [42]
- ROBTET teleoperated system for live-line maintenance [43]
- Robot for automatic washing and brushing of polluted electric insulators [44]

8.8 Problems

1. Discuss robot maintenance-related needs and maintenance types.
2. What are the commonly used tools to maintain a robot?
3. Discuss robot diagnosis and monitoring approaches.
4. What are the most useful guidelines to safeguard robot maintenance personnel?
5. What are the safeguarding methods that can be used during the robot maintenance process?
6. Prove Equation 8.15 by using Equation 8.12.
7. Prove Equation 8.30 with the aid of Equation 8.28.
8. Discuss the following two areas of robotics applications in maintenance and repair:
   a. Nuclear industry
   b. Railways

9. Discuss the following three areas of robotics applications in maintenance and repair:

    a. Highways

    b. Power line maintenance

    c. Underwater facilities

10. List at least 10 tooling items that can be used for periodic robot inspections.

References

1. Dhillon, B.S., *Robot Reliability and Safety*, Springer-Verlag, New York, 1991.
2. Lester, W.A., Lannon, R.P., Bellandi, R., Robot users need to have a program for maintenance, industrial engineering, Part I, *Industrial Engineering*, Vol. 17, 1985, pp. 28–32.
3. Japanese Ministry of Labor, Japanese Industrial Safety and Health Association, ed., *An Interpretation of the Technical Guidance on Safety Standards in the Use, Etc., of Industrial Robots*, Tokyo, 1985.
4. Clark, D.R., Lehto, M.R., Reliability, maintenance, and safety of robots, in *Handbook of Industrial Robotics*, ed. S.Y. Nof, John Wiley and Sons, New York, 1999, pp. 717–753.
5. Robotics Industries Association, *American National Standard for Industrial Robots and Robot Systems: Safety Requirements*, Document No. ANSI/RIA R15.06, 1986, 900 Victors Way, P.O. Box 3724, Ann Arbor, Michigan 48016, 1986.
6. Lodge, J.E., How to protect robot maintenance workers, *National Safety News*, 1984, pp. 48–51.
7. Wild, R., *Essentials of Production and Operations Management*, Holt, Rinehart, and Winston, London, 1985.
8. Dhillon, B.S., *Mechanical Reliability: Theory, Models, and Applications*, American Institute of Aeronautics and Astronautics, Washington, DC, 1988.
9. Dhillon, B.S., *Design Reliability: Fundamentals and Applications*, CRC Press, Boca Raton, Florida, 1999.
10. Pages, A., Gondran, M., *System Reliability*, Springer-Verlag, New York, 1986.
11. Morse, P.M., *Queues, Inventories, and Maintenance*, John Wiley and Sons, New York, 1958.
12. Jardine, A.K.S., *Maintenance, Replacement, and Reliability*, Pitman Publishing, London, 1973.
13. Dhillon, B.S., *Engineering Maintenance: A Modern Approach*, CRC Press, Boca Raton, Florida, 2002.
14. Parker, L.E., Draper, J.V., Maintenance and repair, in *Handbook of Industrial Robotics*, ed. S.Y. Nof, John Wiley and Sons, New York, 1999, pp. 1023–1036.
15. Vertut, J., Coiffet, P., *Teleoperation and Robotics: Applications and Technology*, Prentice-Hall, Englewood Cliffs, 1985.

16. Chesser, J.B., *BRET Rack Remote Maintenance Demonstration Test, Report No. ORNL/TM-10875*, Oak Ridge National Laboratory, Oak Ridge, 1988.
17. Lovett, J.T., Development robotic maintenance systems for nuclear power plants, *Nuclear Plant Journal*, Vol. 9, 1991, pp. 87–90.
18. Macdonald, D. et al., Remote replacement of TF and PF coils for the compact ignition Tokamak, *Proceedings of the 4th ANS Topical Meeting on Robotics and Remote Systems*, 1991, pp. 131–140.
19. Tsukune, H. et al., Research and development of advanced robots for nuclear power plants, *Bulletin of the Electrochemical Laboratory*, Vol. 58, No. 4, 1994, pp. 51–65.
20. Glass, S.W. et al., Modular robotic applications in nuclear power plant maintenance, *Proceedings of the 58th American Power Conference*, 1996, pp. 421–426.
21. Ali, M., Puffer, R., Roman, H., Evaluation of a multifingered robot hand for nuclear power plant operations and maintenance tasks, *Proceedings of the 5th World Conference on Robotics Research*, 1994, pp. 217–227.
22. Horne, R.A. et al., Extended tele-robotic activities at CERN, *Proceedings of the 4th ANS Topical Meeting on Robotics and Remote Systems*, 1991, pp. 525–534.
23. Merchant, D.J., Tarpinian, J.E., Post-accident recovery operations at TMI-2, *Proceedings of the Workshop on Requirements of Mobile Teleoperators for Radiological Emergency Response and Recovery*, 1985, pp. 1–9.
24. Bengal, P.R., The TMI-2 remote technology program, *Proceedings of the Workshop on Requirements of Mobile Teleoperators for Radiological Emergency Response and Recovery*, 1985, pp. 49–60.
25. Chester, C.V., Characterization of radiological emergencies, *Proceedings of the Workshop on Requirements of Mobile Teleoperators for Radiological Emergency Response and Recovery*, 1985, pp. 37–48.
26. Noakes, M.W., Haley, D.C., Willis, W.D., The selective equipment removal system dual arm work module, *Proceedings of the 7th Topical Meeting on Robotics and Remote Systems*, 1997, pp. 478–483.
27. Randolph, J.D. et al., Development of waste dislodging and retrieval system for use in the Oak Ridge National Laboratory Gunnite Tanks, *Proceedings of the 7th Topical Meeting on Robotics and Remote Systems*, 1997, pp. 894–906.
28. Kiebel, G.R., Carteret, B.A., Niebuhr, D.P., Light duty utility arm deployment in Hanford Tank T-106, *Proceedings of the 7th Topical Meeting on Robotics and Remote Systems*, 1997, pp. 921–930.
29. Draper, J.V., Function analysis for the single-shell tank waste retrieval manipulator system, Report No. ORNL/TM-12417, Oak Ridge National Laboratory, Oak Ridge, 1993.
30. Draper, J.V., Task analysis for the single-shell tank waste retrieval manipulator system, Report No. ORNL/TM-12432, Oak Ridge National Laboratory, Oak Ridge, 1993.
31. Ravani, B., West, T., Applications of robotics and automation in highway maintenance operations, *Proceedings of the 2nd International Conference on Applications of Advanced Technologies in Transportation Engineering*, 1991, pp. 61–65.
32. Zhou, T., West, T., Assessment of the state-of-the-art of robotics applications in highway construction and maintenance, *Proceedings of the 2nd International Conference on Applications of Advanced Technologies in Transportation Engineering*, 1991, pp. 56–60.

33. Nease, A., Development of Rapid Runway Repair (RRR) telerobotic construction equipment, *Proceedings of the IEEE National Telesystems Conference*, 1991, pp. 321–322.
34. Martland, C., Analysis of the potential impacts of automation and robotics on locomotive rebuilding, *IEEE Transactions on Engineering Management*, Vol. 34, No. 2, 1987, pp. 92–100.
35. Wiercienski, W., Leek, A., Feasibility of robotic cleaning of undersides of Toronto's subway cars, *Journal of Transportation Engineering*, Vol. 116, No. 3, 1990, pp. 272–279.
36. Toward the "Smart" rail maintenance system, *Railway Track & Structures*, Vol. 82, No. 11, 1986, pp. 21–24.
37. Edahiro, K., Development of "Underwater Robot" cleaner for live growth in power station, in *Teleoperated Robotics in Hostile Environments*, eds. H.L. Martin and D.P. Kuban, Robotics International Society of Manufacturing Engineers, Dearborn, Michigan, 1985, pp. 108–118.
38. Travato, S.A., Ruggieri, S.K., Design, development and field testing of CECIL: A steam generator secondary side maintenance robot, *Proceedings of the 4th ANS Topical Meeting on Robotics and Remote Systems*, 1991, pp. 121–130.
39. Yemington, C., Telerobotics in the offshore oil industry, *Proceedings of the 4th ANS Topical Meeting on Robotics and Remote Systems*, 1991, pp. 441–450.
40. Gallimore, D., Madsen, A., *Remotely Operated Vehicles of the World*, Herfordshire: Oilfield Publications, Ledbury, U.K., 1994.
41. Dunlap, J., Robotic maintenance of overhead transmission lines, *IEEE Transactions on Power Delivery*, Vol. 1, No. 3, 1986, pp. 280–284.
42. Yano, K. et al., Development of the semi-automatic hot-line work robot system "Phase II," *Proceedings of the 7th International Conference on Transportation and Distribution Construction and Live Line Maintenance*, 1995, pp. 213–218.
43. Aracil, R. et al., ROBTET: A new teleoperated system for live-line maintenance, *Proceedings of the 7th International Conference on Transmission and Distribution Construction and Live-Line Maintenance*, 1995, pp. 205–211.
44. Yi, H., Jiansheng, S., The research of the automatic washing-brushing robot of 500 kv DC insulator string, *Proceedings of the 6th International Conference on Transmission and Distribution Construction and Live Line Maintenance*, 1993, pp. 411–424.

9

Human Factors and Safety Considerations in Robotics Workplaces

9.1 Introduction

Human factors and safety are important in the design of robotic workplaces, as they are intimately related. More clearly, many safety problems have solutions in human factors.

Generally, in regard to robotics, human factors are basically concerned with the way robotic equipment interface systems and the environment should be designed so that they are effectively compatible with the humans who use it. In industrial robotics, in addition to ambient conditions, basic concern is associated with the design of items such as controller panels, teach pendants, and computer terminals [1].

Needless to say, one of the major motives for investing in industrial robotics is to relieve human operators from tasks that are hazardous or difficult. Over the years, significant progress has been made in the area of human factors and safety considerations in robotics workplaces. This chapter presents various important aspects of human factors and safety considerations in robotics workplaces.

9.2 Human Factors-Related Issues during the Robotic Systems' Factory Integration Process

During the factory integration of robotic systems, there are many issues related to human factors that must be taken into consideration with care. The primary eight of these issues are the following [2,3]:

- Safety
- Maintainability
- Worker–machine interface

- Selection and training
- Work environment
- Management
- Job design
- Communication among workers and between workers and management

Each of the above issues is described below, separately.

9.2.1 Safety

The safety of workers and other individuals is a very serious issue and must be considered with care during the factory integration process of robotic systems. Three aspects to robot-related safety are as follows [2,3]:

- Providing appropriate protection to robot operators, maintenance personnel, and programmers
- Providing appropriate protection to the robots themselves
- Providing safety to curious outsiders

Furthermore, there are several issues that require careful attention. These include designing an effective production layout with regard to safety, reducing the effect of environmental hazards, and labeling potential hazards.

9.2.2 Maintainability

Robot maintainability is an important issue, as diagnostic-related activities can account for around 70% of maintenance time. Thus, some of the useful human factors-related guidelines/notes for the maintenance of robotic systems are presented below [2,3]:

- Ensure that robotic system maintenance is performed by personnel who work in the same environment as the robots.
- Maintenance personnel require appropriate multicraft capability as well as skills in robot system operation and programming knowledge for carrying out diagnoses and tests.
- The response time of maintenance personnel to a robotic workstation could be affected by a large number of factors, including craft coordination, the availability of qualified personnel, and the inclusion of robots on a priority list for maintenance.
- Appropriate maintenance manuals are required to carry out robotic maintenance effectively.

- Various specialties are needed for robotic maintenance (e.g., oiling, mechanical, electronic, and plumbing).
- Ensure that well-qualified maintenance personnel are provided in all operating shifts to have effective production results.

9.2.3 Worker–Machine Interface

The main objective of the worker–machine interface issue is to design equipment to match capabilities and shortcomings of concerned workers effectively. In regard to robotics, as robot systems are maintained, operated, or programmed by humans, they are man–machine systems.

Two useful worker–machine interface guidelines concerning the factory integration of robotics are as follows:

- During the interface design phase, consider with care items such as the tasks of the robot maintenance personnel, operators, and programmers, and the location of the controls.
- For each and every robot application and man–machine interactive task, determine the most suitable speed range for the robot. In addition, aim to make the robot speed (i.e., pace rate) be operator-paced.

9.2.4 Selection and Training

The issue of selection and training is concerned with the robot maintenance personnel, operators, and programmers. It involves selecting and locating appropriate workers, placing the workers appropriately for the relevant job, and training or retraining those workers to have the necessary skill, knowledge, and ability.

Finally, an effective training program requires careful consideration as to the training objective, the training method, the evaluation approaches, and the needs of the job or task.

9.2.5 Work Environment

The work environment incorporates both the physical and psychological environments under which a worker concerned with robotic systems performs his/her task; this includes issues of the potential for experiencing discomfort in the given environment. The physical environment includes items such as glare, noise, low ambient illumination, and an incorrect work position.

Alongside, the psychological environment includes items such as improperly devised rest periods, disagreeable coworkers, and monotonous and hectic tasks.

9.2.6 Management

Issues of management include elements such as the following:

- The effect of management decisions on the morale and motivation of workers concerned with robotics systems
- Employee relations
- Organizational design

9.2.7 Job Design

The issue of job design is a quite challenging human factor issue. The objective is to design jobs/tasks such that the individual differences are clearly recognized, in addition to the maximization of individual productivity and satisfaction.

Here, it is to be noted that the performance of task analysis is very important during designing jobs.

9.2.8 Communication among Workers and between Workers and Management

The issue of communications is concerned with establishing effective formal and informal communication channels among various groups of people, for example, between robot maintenance personnel and operators, between robot operators on different shifts, and between robot programmers and operators.

It is to be noted with care that effective communication among workers and between workers and management personnel is very important for the successful factory integration of robotic systems.

9.3 Common Robot and Robot-Related Human Tasks

Robots are used to perform various types of tasks. The common ones are shown in Figure 9.1 [4].

Similarly, the common robot-related human tasks are as follows [4]:

- *Maintenance.* This human task is concerned with maintenance of robots; it is complex and generally requires knowledge of mechanics, electronics, hydraulics, and computers.
- *Programming.* This human task is concerned with programming of robots, commonly involving the following:
  - By using a teach pendant
  - By manually leading the robot arm
  - Through a computer terminal

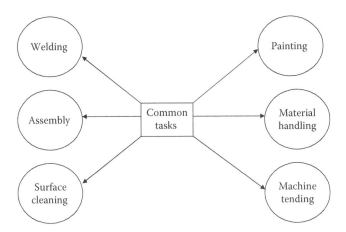

FIGURE 9.1
Common robot tasks.

- *Monitoring/supervising.* This robot-related human task involves the detection of events that require operator attention. In this case, the operator has two options in monitoring the situation: directly monitoring by observing the robot or indirectly monitoring by observing displays.
- *Intervening/helping.* This human task is concerned with intervening/ helping, as robots often require help to carry out complex tasks. For example, in robotic assembly, gravity-feed devices have to be cleared and items that the robot drops have to be picked up from the floor; thus, in this case, the robot operator is helping/serving the robot.
- *Leftover tasks.* These human tasks are those that the robot has difficulty in performing. One example of such tasks is visual inspection and pattern recognition.
- *Replace robot during failure.* This human task is concerned with replacing the robot during failure in performance as sometimes it becomes essential for operators to replace robots.

9.4 Rules of Robotics in Regard to Humans and Advantages and Disadvantages of Robotization with Respect to Human Factors

Over the years, many rules of robotics with regard to humans have been established. Seven of these rules are as follows [3,5]:

- Robots must be easily operable by all involved humans.

- Robots must only replace humans on jobs/tasks that are considered to be hazardous or which the human beings are clearly unwilling to undertake.
- Robots must not psychologically and physically oppress human beings.
- Robots must be used and produced with the objective of contributing to the welfare and development of human beings.
- Whenever robots replace humans in certain jobs/tasks, obtain the prior approval of those who will be affected.
- All robots must be under the command of human beings, so that other human beings are not harmed and only robots themselves are damaged.
- At the accomplishment of assigned tasks or jobs, robots must leave the area to avoid interference with humans and other robots in the area.

There are many advantages and disadvantages of robotization, with respect to human factors. Some of the advantages are as follows [3,6]:

- To free workers from having to carry out tasks involving a heavy physical load
- To free workers from carrying out difficult and monotonous jobs/tasks
- To free workers from having to carry out jobs/tasks in the presence of high noise, vibrations, and so on.
- To free workers from carrying out jobs/tasks at high temperatures
- To free workers from having to carry out microscopic tasks/jobs that need acute eyesight
- To free workers from having to carry out tasks/jobs in the environment of harmful gases
- To supplement the capabilities of handicapped and old persons
- To free workers from carrying out jobs/tasks exposed to nuclear radiation

Similarly, some of the disadvantages of robotization with respect to human factors are as follows [3,6]:

- Human safety-related problems
- Difficult to operate because of significant variations of control panel layouts from one root model to another

- Incompatibility in terms of motion with humans— for example, robots move at acute angles or linearly and abruptly stop but humans do not
- Incompatibility with humans in regard to work pace

9.5 Humans at Risk from Robots and Risk-Reducing Measures to Prevent Robot-Related Human Accidents

Experiences over the years clearly indicate that there are several people who could be at risk from robots used in assembly-oriented work. These people are categorized under the following four groups [3,7]:

- *Robot programmers.* These people are subject to various types of injuries because they come into direct physical contact with robots. Although, during the programming process, robots move just a small fraction of their normal operating speeds, failures do occur from time to time and programmers may get injured.
- *Maintenance personnel.* These people are in direct physical contact with robots because of the nature of their work. Thus, they could be subject to hand or face injuries, electrocution, and so on.
- *People outside the identified danger zone.* These individuals are also at risk to a certain degree because, under certain circumstances, the robot gripper may fling out parts/components that may, in turn, strike them.
- *Casual observers.* These individuals are also at risk from injury by robots because of their ignorance. For example, they may assume that the robot is in a stationary state when, in fact, it is not.

Some of the risk-reducing measures to prevent robot-related human accidents are as follows [4,8]:

- *Measure I*: Paint the robot with a special color for making its position sharply visible
- *Measure II*: Equip the robot with a touch-sensitive skin so that it automatically shuts down in the event of a physical contact with a human
- *Measure III*: Locate the robot so that all the pinch-points are totally eliminated
- *Measure IV*: Design the robot to lower or eliminate the risk of cuts and pressure contusions

9.6 Useful Guidelines to Safeguard Robot Teachers and Operators

The safety of robot teachers is an important issue and it requires a careful consideration. Some of the useful guidelines for safeguarding robot teachers are as follows [3,5]:

- Each robot system user must ensure that all involved robot teachers are properly trained in regard to the recommended "teach" procedures, the control program, and the specific installation
- Require robot teachers to vacate the restricted robot work zone prior to switching to automatic mode
- Function test the teach control of the pendant and repair any damage or failure prior to commencement of the teaching operation
- At the time of choosing the teach mode, fully satisfy conditions such as cutting of the robot from responding to any signals that may cause motion, putting complete control of any other equipment in the prohibited work area in the hands of the robot teacher(s) if their movements would present a potential hazard, and putting the complete robot system under the control of the teacher(s)
- Require robot teachers to properly inspect the robot visually, including its restricted work area, before starting teaching (the robot), in order to ensure that hazard-causing conditions do not exist
- Require robot teachers to ensure that all safeguards are properly in place and functional, as specified, in the teach mode before entry into the prohibited robot work area

Some of the useful guidelines for safeguarding robot operators are as follows [3,5]:

- Provide appropriate safeguards to prevent the entry of robot operators into the prohibited work area while the robot is in motion
- Each robot system user must establish appropriate safeguards for each and every operation associated with the robot system
- Provide appropriate training to robot system operators in the proper operation of the robot system's control actuators
- Provide appropriate training to robot system operators so that they can quickly recognize known hazards associated with each and every task involving the robot system

- Provide appropriate training to robot system operators so that they can respond properly and quickly to recognized hazardous conditions

9.7 Approaches for Limiting Robot Movements

There are many approaches that can be used for limitation of robot movements. Six of these approaches are shown in Figure 9.2 [4].

Motion-limiting devices are used to restrict movements of robots by using limiting switches and fixed stops located close to the robot joints. Robot-locking devices such as chains and pins are used to keep the robot in one location in situations when the robot is inactive or has been temporarily halted, due to, say, maintenance work.

Brakes are applied to the joints of the robot, which may stop the movement of robot at any location. An important criterion to evaluate the brakes' effectiveness is the stopping distance. The "watchdog safety computer" is a stand-alone microcomputer system used for monitoring movements of robots in their workstations; its purpose is to detect operations that are outside the normal conditions' range as well as to stop robots before collision or damage.

Manual stops include emergency stop buttons and advance switches; speed limitation for teaching and maintenance approach is considered self-explanatory.

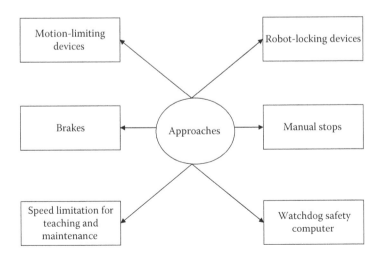

FIGURE 9.2
Approaches for limiting robot movement.

9.8 Methods for Analysis of Safety and Human Error in Robotics Workplaces

There are many methods developed over the years that can be used to analyze safety and/or human error in robotics workplaces [4,9]. A number of these methods are described below, separately.

9.8.1 Deviation Analysis

This method is concerned with analyzing accidents that result from deviations or disturbances in the planned or normal production process. These deviations are applicable to organizational, technical, and individual functions. The method is composed of the following three steps [4,10]:

- *Step 1*: This step is concerned with summarizing the system functions and different activities involved, and then dividing the activities into blocks of suitable size. Furthermore, for each and every block, the normal activities and possible deviations from normal activities are listed.
- *Step 2*: This step is concerned with judging the consequences of the deviations, such as dangerous situations or conditions.
- *Step 3*: This step is concerned with determining the systems improvements that would lower the severity levels of problematic consequences.

9.8.2 Energy Barrier Analysis

The basis for this method is that accidents occur due to release or transfer of energy that becomes uncontrollable. The method is composed of the following four steps [4,11]:

- *Step 1*: Highlight all possible sources of potential unwanted energy (PUE) in the robot-associated environment
- *Step 2*: Highlight effective mechanisms for lowering and controlling PUE levels from the source
- *Step 3*: Highlight effective mechanisms to block the transfer of PUE
- *Step 4*: Highlight effective mechanisms to warn the operator of the existence of PUE

It is to be noted that PUE is generally generated from environmental products that the robot is operating with (i.e., chemical, thermal, etc.), or from the robot (hydraulic or electrical power) itself. As per [11], the most practical barrier solution approach may be to block the transfer of PUE to the human operator. This may imply making use of devices such as motion sensors, vision sensors, and floor mats, all of which must be sensitive to human proximity.

9.8.3 Task Analysis

This method is concerned with breaking down the main task into subtasks and then into task elements. These may then be analyzed in regard to information processing, information input and output. In situations where there is cooperation between humans and robots, it may be quite appropriate to carry out a dual task analysis of the human-related and robot-related actions in combination [4]. A mathematical approach for modeling the reliability of a system, where actions are executed intermittently by the human and the robot, is available in [12].

Additional information on this method is available in [4,13].

9.8.4 Fault Tree Analysis

This is a very useful method and, over the years, has been used to analyze the safety and reliability of human–machine systems. In 1983, National Institute for Occupational Safety and Health (NIOSH) proposed this method to analyze maintenance aspects of industrial machines [14].

As per past experiences, since maintenance is the primary activity that has been implicated in the occurrence of robot accidents, fault tree analysis is considered highly relevant [4]. The method is described in Chapter 4 and in extensive detail in [15–16]. The application of the fault tree method to robot accidents in workplaces is demonstrated through the simple example presented below [3].

EXAMPLE 9.1

Assume that in a robotics workplace a robot accident involving a human can only occur due to two reasons: human in the robotics work zone and a sudden movement of the robot. The presence of the human in the robotics work zone could be either due to intentional entry or unintentional entry. Similarly, the sudden movement of the robot(s) can happen due to two reasons: power supply on and unexpected start signal.

Develop a fault tree for the top event: robot accident involving a human and then calculate the probability of occurrence of the top event if the probabilities of occurrence of events (i.e., reasons) are, respectively, 0.05, 0.15, 0.2, and 0.1, for the power supply being on, an unexpected start signal, intentional entry, and unintentional entry.

By using the fault tree symbols given in Chapter 4, a fault tree for the example is shown in Figure 9.3. The single capital letters in the figure denote corresponding fault events (i.e., *T*: robot accident involving a human, *A*: sudden movement of robot, *B*: human in the robotics work zone, *C*: power supply on, *D*: unexpected start signal, *E*: intentional entry, *F*: unintentional entry).

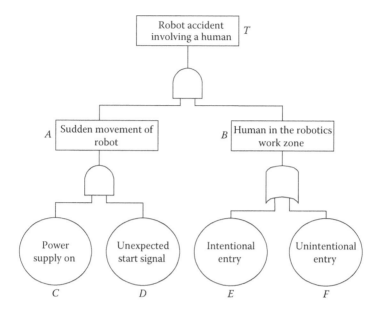

FIGURE 9.3
A fault tree for Example 9.1.

With the aid of Chapter 4, material concerning fault tree analysis and the given data values, the probability of occurrence of event A is

$$P(A) = P(C)P(D)$$
$$= (0.05)(0.15)$$
$$= 0.0075 \qquad (9.1)$$

where
$P(C)$ is the probability of occurrence of event C.
$P(D)$ is the probability of occurrence of event D.

Similarly, with the aid of Chapter 4, material concerning fault tree analysis and the given data values, the probability of occurrence of event B is

$$P(B) = P(E) + P(F) - P(E)P(F)$$
$$= 0.2 + 0.1 - (0.2)(0.1)$$
$$= 0.28 \qquad (9.2)$$

where
$P(E)$ is the probability of occurrence of event E.
$P(F)$ is the probability of occurrence of event F.

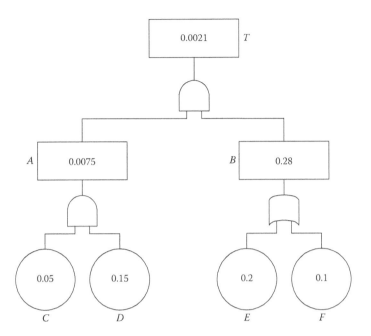

FIGURE 9.4
A fault tree with probability values for the given and calculated fault event occurrence.

Using the above-calculated values and Chapter 4 material concerning fault tree analysis, we get

$$P(T) = P(A)\, P(B)$$
$$= (0.0075)\,(0.28)$$
$$= 0.0021 \tag{9.3}$$

where
$P(T)$ is the probability of occurrence of event T.

Thus, the probability of occurrence of the top event T (i.e., a robot accident involving a human) is 0.0021. The fault tree in Figure 9.3 with given and calculated probability values for fault event occurrence is shown in Figure 9.4.

9.8.5 Markov Method

This is a quite useful method and, over the years, has been used to perform analyses of safety and reliability of human–machine systems [17]. The method is described in Chapter 4 and its application to various types of human–machine systems are demonstrated in [17].

The application of the Markov method to robot-related accidents in work-places is demonstrated through the example below.

EXAMPLE 9.2

Assume that a robot used in a workplace can either fail due to an accident involving a human or fail safely. The robot failure rate due to an accident involving a human is λ_a and its safe failure rate is λ_s. The state-space diagram describing this scenario is shown in Figure 9.5. The numerals in the circles and box denote robot states.

Develop expressions for the robot state probabilities by using the Markov method and assuming that the robot failures occur independently and its failure rates are constant.

The following symbols are associated with the Figure 9.5 state-space diagram:

   i is the ith state of the robot, where $i = 0$ (robot operating normally), $i = 1$ (robot failed due to an accident involving a human), $i = 2$ (robot failed safely).
   $P_i(t)$ is the probability that the robot is in state i at time t, for $i = 0,1,2$.
   λ_s is the constant failure rate of the robot failing safely.
   λ_a is the constant failure rate of the robot failing due to an accident involving a human.

Using the Markov method described in Chapter 4 and Figure 9.5, we write down the following equations:

$$\frac{dP_0(t)}{dt} + (\lambda_a + \lambda_s)P_0(t) = 0 \tag{9.4}$$

$$\frac{dP_1(t)}{dt} - \lambda_a P_1(t) = 0 \tag{9.5}$$

$$\frac{dP_2(t)}{dt} - \lambda_s P_2(t) = 0 \tag{9.6}$$

At time $t = 0$, $P_0(0) = 1$, $P_1(0) = 0$, and $P_2(0) = 0$.

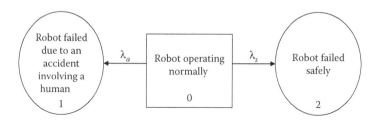

FIGURE 9.5
State-space diagram of a robot used in a workplace.

By solving Equations 9.4 through 9.6, we get

$$P_0(t) = e^{-(\lambda_a + \lambda_s)t} \tag{9.7}$$

$$P_1(t) = \frac{\lambda_a}{\lambda_a + \lambda_s}[1 - e^{-(\lambda_a + \lambda_s)t}] \tag{9.8}$$

$$P_2(t) = \frac{\lambda_s}{\lambda_a + \lambda_s}[1 - e^{-(\lambda_a + \lambda_s)t}] \tag{9.9}$$

EXAMPLE 9.3

Assume that the constant failure rate of a robot used in a workplace and failing safely is 0.05 failures per hour and its constant failure rate failing due to an accident involving a human is 0.01 failures per hour. Calculate with the aid of Equations 9.7 through 9.9, the probabilities of the robot being in states 0, 1, and 2 during a 100 h mission.

By substituting the given data values into Equations 9.7 through 9.9, we obtain

$$P_0(100) = e^{-(0.01+0.05)(100)}$$

$$= 0.0024$$

$$P_1(100) = \frac{0.01}{(0.01 + 0.05)}[1 - e^{(0.01+0.05)(100)}]$$

$$= 0.1662$$

and

$$P_2(100) = \frac{0.05}{(0.01 + 0.05)}[1 - e^{-(0.01+0.05)(100)}]$$

$$= 0.8313$$

Thus, the probabilities of the robot being in states 0, 1, and 2 during a 100 h mission are 0.0024, 0.1662, and 0.8313, respectively.

9.9 Problems

1. What are the primary human factors-related issues during the robotic systems' factory integration process?
2. Discuss in detail four issues of the above question.

3. What are the common robot tasks?

4. List and discuss common robot-related human tasks.

5. What are the rules of robotics with regard to humans?

6. What are the advantages and disadvantages of robotization with respect to humans?

7. What are the risk-reducing measures to prevent robot-related human accidents?

8. Discuss at least six useful guidelines to safeguard robot teachers.

9. What are the robot movement limitation approaches? Discuss each of these approaches in detail.

10. Describe the following methods with regard to robotics workplaces:

    a. Energy barrier analysis

    b. Deviation analysis

    c. Fault tree analysis

References

1. Parsons, H.M., Human factors loom as vital issues in robotics, *Robotics Today*, 1987, pp. 27–31.

2. Howard, J.M., Focus on the human factors in applying robotic systems, *Robotics Today*, 1982, pp. 32–34.

3. Dhillon, B.S., *Robot Reliability and Safety*, Springer-Verlag, New York, 1991.

4. Helander, M.G., Ergonomics and safety considerations in the design of robotics workplaces: A review and some priorities for research, *International Journal of Industrial Ergonomics*, Vol. 6, 1990, pp. 127–149.

5. American National Standard for Industrial Robots and Robot Systems—Safety Requirements, *Document No. ANSI/RIA R15.06*, 1986; available from the Robotic Industries Association (RIA), 900 Victors Way, P.O. Box 3724, Ann Arbor, Michigan.

6. Yokomizo, Y., Hasegawa, Y., Komatsubara, A., Problems of, and industrial medicine measures for, the introduction of robots, *Occupational Health and Safety in Automation and Robotics* ed. K. Noro, Taylor & Francis, London, 1987, pp. 309–325.

7. Owen, T., *Assembly with Robots*, Kogan Page, London, 1985.

8. Carlsson, J., *Robot Accidents in Sweden, Report No. 1984.2*, Arbetarskyddsstgrelsen, Stockholm, Sweden, 1984.

9. Dhillon, B.S., *Engineering Safety: Fundamentals, Techniques, and Applications*, World Scientific Publishing, River Edge, New Jersey, 2003.

10. Harms-Ringdahl, L., Experiences from safety analysis of automatic equipment, *Journal of Occupational Accidents*, Vol. 8, 1986, pp. 139–148.

11. Rahimi, M., Systems safety for robots: An energy barrier analysis, *Journal of Occupational Accidents*, Vol. 8, 1986, pp. 127–138.

12. Umezaki, S., Sugimoto, N., A study on safety evaluation index and industrial accident analysis from the viewpoint of the safety confirmation type, in *Ergonomics of Hybrid Automated Systems I*, ed. H.R. Parsaei and M.R. Wilhelm, Elsevier, Amsterdam, The Netherland, 1988, pp. 491–498.

13. Kirwan, B., Ainsworth, L.K., *A Guide to Task Analysis*, Taylor & Francis, London, 1992.

14. *Guidelines for Controlling Hazardous Energy*, National Institute for Occupational Safety and Health (NIOSH), Morgan Town, West Virginia, 1983.

15. Dhillon, B.S., *Design Reliability: Fundamentals and Applications*, CRC Press, Boca Raton, Florida, 1999.

16. Dhillon, B.S., Singh, C., *Engineering Reliability: New Techniques and Applications*, John Wiley and Sons, New York, 1981.

17. Dhillon, B.S., *Human Reliability: With Human Factors*, Pergamon Press, New York, 1986.

10

Robot Testing, Costing, and Failure Data

10.1 Introduction

Robot testing is very important because it plays an important role in the success or failure of robotization. Two common types of robot testing are known as robot performance testing and robot reliability testing [1]. Over the years, a number of methods have been developed to conduct performance testing; the forms of performance testing vary from the verification of design goals by robot manufacturers to the determination of the most suitable robot for a specific application. There are various types of robot reliability testing and their purpose is to obtain knowledge of robot failures.

Robot costing is very important as cost plays a crucial role, directly or indirectly, in robotization in the industrial sector. For example, the economic viability of robot use in the factory depends on factors such as investment cost (i.e., acquisition cost, installation cost, and cost of adaptation to the workshop), ownership cost (e.g., maintenance cost and fuel cost), and return on investment [2].

Robot failure data provides invaluable information to concerned professionals and management alike as it indicates the results of the reliability-related effort put in during the design and manufacture phases of robots. Today, there are various sources for obtaining data related to robot failures.

This chapter presents various important aspects of robot testing, costing, and failure data.

10.2 Robot Performance Testing

There are several forms of robot performance testing. Two examples of such forms are the verification of design goals by the robot manufacturer and the determination of the most suitable robot for a specific application. Three performance parameters that generally play a key role in robot performance testing are shown in Figure 10.1 [1,3].

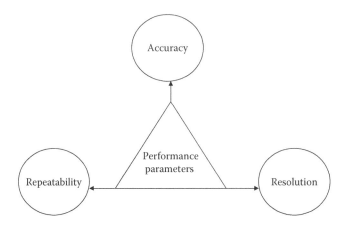

FIGURE 10.1
Performance parameters that play a key role in robot performance testing.

Accuracy is concerned with measuring the differential between the directions given to a robot regarding movement(s) and its actual movements. Repeatability may simply be described as the closeness of concurrence of a robot's repeated movements in the same location when subject to the same conditions [1,4]. It is to be noted that repeatability is one of the greatest advantages of a robot because of its ability to repeat programmed motions consistently.

Finally, resolution, in regard to robots, is determined by the capability of feedback devices such as encoders and resolvers for determining the locality of a specific point and, ultimately, the end point position of the robot arm. Nonetheless, in broader perspectives, resolution may simply be stated as a measure of the change in variable output of a device.

It is to be noted that most of the industrial robot performance tests may be tailored to measure either of two types of behavior: path and point-to-point. Parameters such as kinematic repeatability or kinematic accuracy are measured by path tests as opposed to dynamic stability, quasi-static repeatability, or quasi-static accuracy by point-to-point tests.

Some of the general guidelines concerned with industrial robot performance testing are presented below [1,5]:

- Ensure that the appropriate safety rules are properly followed during the robot testing and programming process.
- Ensure that the involved test equipment is temperature- and time-stable.
- Ensure that for tests with multiple permutations, only one control parameter at a time is changed, because parameters such as repeatability depend upon the robot's speed, payload, temperature, and so on.

- Ensure the proper rigidity of the mounting platform of the robot.
- Ensure that noncontact gauges such as laser interferometers, analog distance sensors, and vision systems are employed instead of contact gauges (e.g., dial indicators).
- Ensure that the measuring equipment resolution is at least one order of magnitude greater than the actual value to be measured.

10.3 Robot Performance Testing Methods

Over the years, many test methods have been developed to evaluate the performance of robots [5]. Four of these test methods are described below, separately.

10.3.1 Test-Robot Program Method

The test-robot program method is concerned with measuring the position and orientation errors acquired during static positioning, and are computed statistically as three-dimensional error vectors. The method has two versions (i.e., Mark I and Mark II) developed by the same individual [5,6]. Some of the important features of this method are as follows [1,5,6]:

- It possesses the capability to be employed as a simulation tool to evaluate alternative solutions.
- It possesses the user-friendly and run-time-checked screen management system.
- It possesses the capability to undertake analysis for hand position and orientation error for industrial robots.
- It possesses the capability to save end results in a user-specified text file on disk.
- It possesses the capability for evaluating perpendicularity errors of the used test rig for robots.
- It possesses the capability for disk output/input error handling and run-time input line editing.

10.3.2 Ford Method

The Ford method was developed by the Ford Motor Company. It describes industry-oriented approaches to certify industrial robots using the Selspot, Inc., vision system [1]. The method calls for putting the robot in question under a cycling test consisting of the movement of all axes from one end of motion to the other at full-rated payload and at speed for about 40 h continuously.

This is basically a noninstrumented test and any duty cycle-related limitation is followed closely. The two objectives of the test are as follows [1]:

- Obtain information concerning the integrity and reliability of the robot system
- Find infant mortality faults such as usage of undersized fuses and loose bolts

Additional information on this method is available in [1].

10.3.3 IPA–Stuttgard Method

The IPA–Stuttgard method illustrates a general procedure to measure static positioning errors and dynamic path parameters. It was developed at the Institute for Production and Automation (IPA), Stuttgart, Germany [1,5,7]. Two main requirements considered during the development of this robot performance testing method are as follows [1,7]:

- *Requirement I: Highly automated test bed.* The main idea behind this requirement is to lower the time required to evaluate the performance of a robot.
- *Requirement II: Exact coordination between the measuring equipment used and the robot.* The main reason behind this requirement is to permit the measurement of characteristics within the work zone of a robot with an acceptable level of accuracy.

Additional information on this method is available in [7].

10.3.4 National Bureau of Standards Method

The National Bureau of Standards (NBS) of the United States developed this robot performance test method to measure the performance of industrial robots. The NBS method uses laser-based instrumentation [1,8]. The basis for the method is the idea of continuously controlling the angles of projection of a laser beam, so that the beam will impinge on the target mirror installed on the robot's wrist.

Additional information on the method is available in [5].

10.4 Robot Reliability Testing

The basic objective of robot reliability testing is to gain knowledge about robot failures, as this knowledge plays an important role in controlling

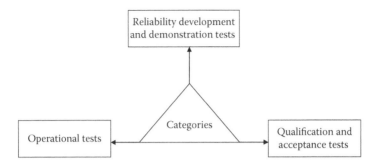

FIGURE 10.2
Categories of robot reliability tests.

failure tendencies and their effects. Robot reliability tests may be grouped under three broad categories as shown in Figure 10.2 [1,9].

The main objectives of the reliability development tests and their demonstration are to verify improvements in the reliability of the design, to determine if the robot design has to be improved to meet the specified reliability requirements, and to point out any required design changes. Two objectives of the qualification and acceptance tests are to determine if a specific design is qualified for its proposed use and to determine if an item is acceptable or not acceptable.

Finally, the objectives of operational tests are to verify the reliability analyses performed during the design stage, to provide the necessary data to justify required operational procedure and operational changes with regard to robot reliability and maintainability, and to provide information that would be useful in the activities to be followed.

10.4.1 Management-Related Tasks in a Robot Reliability Test Plan

A typical robot test plan is composed of many tasks that generally require frequent updating. The plan includes tasks such as the following [1,9]:

- Allocate time and money.
- Determine test needs and objectives.
- Review current or planned functional tests and operations from the standpoint of determining if certain required information could be extracted from such tests and operations.
- Formulate individual test needs.
- Evaluate the already available data for meeting any existing needs, thus eliminating the requirements for specific tests.
- Determine the types of tests to be performed.
- Review and approve test reporting methods and documents.
- Keep information on status of the test.

10.4.2 Useful Guidelines for Robot Reliability Demonstration Tests

Some of the useful guidelines to perform robot reliability demonstration tests are as follows [1,9]:

- Determine the test duration by considering factors such as the availability of the robot equipment to be tested, test facilities, and testing manpower.
- Develop a comprehensive description of the robot equipment with respect to factors such as operational environment, mission time, service life, and operational mission.
- Prepare a list of the robot system items that are to be excluded from the demonstration test under consideration.
- Pay attention to developing appropriate test monitoring approaches whether manual or automatic.
- Put in an appropriate amount of effort to relate test environments to the expected robot use environments.
- Pay attention to choosing the methods to conduct data analysis.
- Pay attention to choosing the test plan.

10.5 Robot Testing and Start-Up Safety-Related Factors

During robot testing and start-up, there are many safety-related factors that have to be considered to improve robot safety. Some of these factors are as follows [1,10]:

- Following the instructions of robot manufacturers for testing and repair.
- After each robot hardware or software modification, maintenance, and repair, a procedure for the restart of the robot should incorporate steps such as function testing the robot for the correct system operation and checking, prior to turning on the power, for any changes or additions to the hardware of the robot.
- Ensuring that no human is present in the work zone of the robot until the safeguarding and the appropriate operation verification is completed.
- The proper incorporation of two verification procedures—that is, I and II. Verification procedure I is concerned with the verification of the installation of various items as required, before turning on the power. Some examples of these items are electrical

connections, limiting devices to restrict the work zone, peripheral equipment and systems, utility connections, mechanical mounting and stability, and communication connections. Similarly, verification procedure II, after turning on the power, is concerned with the verification of items such as the proper functioning of safeguards, interlocks, emergency stop devices, automatic operations as specified, movement of each axis and its restriction as intended, slow speed as specified in [10], and program execution as specified.

10.6 Robot Project Cost

The project cost of robotization may be divided into the following eight categories [1,11]:

- *Category I: Computer cost.* This comprises the hardware and software costs. The hardware cost is associated with items such as the central processor unit, interface modules, and core store. The software cost is associated with items such as customized software and standard software packages.

- *Category II: Safety cost.* This cost accounts for around 10% of the basic system cost and is composed of elements such as the cost of protection devices for the operator(s), of alarm devices, safety enclosure, and safety training.

- *Category III: Manufacturing equipment cost.* This cost is associated with items such as the robot system (this includes the manipulator and associated control system and motive source), grippers or work-handlers, system interfaces, and package enhancements.

- *Category IV: Project management cost.* This cost is associated with managing the project and includes items such as the salary of the project manager and cost of secretarial help.

- *Category V: System engineering cost.* In this cost, labor is the dominant factor because the time of technicians and engineers is quite costly. Labor time is associated with tasks such as purchasing equipment for the system, testing equipment against specifications of manufacturers, testing the system against system criteria, reassembly, and retesting.

- *Category VI: Ancillary equipment and structural modification cost.* This cost is concerned with the work to be carried out prior to the robot installation.

- *Category VII: Cost associated with personnel.* This cost is associated with items such as skills training, redundancies, redeployment, recruitment, and wage settlements.
- *Category VIII: Miscellaneous cost.* This cost includes the costs of items such as feasibility studies, purchase of spares, and maintenance costs.

10.7 Models for Estimating Robot-Related Costs

Over the years, many mathematical models have been developed to estimate various types of robot-related costs. Some of these models are presented below [1].

10.7.1 Model I

This model is concerned with estimating the robot operating cost. The cost is defined by

$$ROC = DLC + RMC + SPFC + MC + ATCC \tag{10.1}$$

where
  ROC is the robot operating cost.
  DLC is the direct labor cost to tend the robot.
  RMC is the cost associated with the robot related maintenance activities.
  $SPFC$ is the cost associated with supplying parts to feeders.
  MC is the miscellaneous cost.
  $ATCC$ is the cost of adjusting tools for changeovers.

10.7.2 Model II

This model is concerned with estimating the investment cost associated with a robotization project. The cost is expressed by

$$RPIC = RC_p + RC_e + RC_{st} + RC_i + RC_m \tag{10.2}$$

where
  $RPIC$ is the investment cost of the robotization project.
  RC_p is the robot procurement cost.
  RC_e is the robot engineering cost.
  RC_{st} is the special tooling cost of the robot.

RC_i is the robot installation cost.

RC_m is the miscellaneous cost of the robotization project.

10.7.3 Model III

This model is concerned with estimating the annual corrective maintenance cost of a robot. The cost is defined by

$$RCMC_a = (LC_m)(OH_{as})\left[\frac{MTTR_r}{MTBF_r}\right] \tag{10.3}$$

where

$RCMC_a$ is the annual corrective maintenance cost of the robot.

OH_{as} is the annual scheduled operating hours of the robot.

LC_m is the robot maintenance labor cost per hour.

$MTTR_r$ is the mean time to repair of the robot.

$MTBF_r$ is the mean time between robot failures.

EXAMPLE 10.1

Assume that the annual scheduled operating hours of a robot are 4000 h and its mean time between failures is 1200 h. The mean time to repair of the robot is 8 h and maintenance labor cost is US$30 per hour.

Calculate the annual corrective maintenance cost of the robot.

By substituting the specified data values into Equation 10.3, we get

$$RCMC_a = (30)(4000)\left[\frac{8}{1200}\right]$$

$$= \$800$$

Thus, the annual corrective maintenance cost of the robot is US$800.

10.7.4 Model IV

This model is concerned with estimating the unit cost of assembly by the robot. The cost is expressed by [12]

$$C_u = T(C_1 + C_2) \tag{10.4}$$

where

C_u is the unit cost of assembly by the robot.

T is the time taken to assemble the product expressed in seconds.

C_1 is the per-second cost of the flexible portions of the flexible assembly system.

C_2 is the per-second cost of the dedicated portions of the flexible assembly system.

10.7.5 Model V

This model is concerned with estimating the possible cost of losses to a robot manufacturer from a robot accident. The cost is expressed by

$$PC_{rf} = C_{ai} + C_{pc} + C_{pm} + C_s + C_{pd} + C_{ii} + C_{lf} + C_m \qquad (10.5)$$

where

PC_{rf} is the possible cost of losses to a robot manufacturer from a robot-related accident.

C_{ai} is the cost of accident investigation.

C_{pc} is the cost of loss in public confidence.

C_{pm} is the cost associated with preventive measures to stop recurrences.

C_s is the settlement cost of death or injury claims.

C_{pd} is the cost associated with property damage claims not covered by insurance.

C_{ii} is the cost of increment in insurance.

C_{lf} is the cost of legal fees for defense against claims.

C_m is the miscellaneous cost.

10.7.6 Model VI

This model is concerned with estimating the robotic installation and start-up cost. The cost is expressed by [13]

$$RISC = AC + PC + TC + EC + IC + ACCS + REC + TRC \qquad (10.6)$$

where

$RISC$ is the robotic installation and start-up cost.

AC is the acquisition cost, which includes costs of the robot and controller.

PC is the programming cost.

TC is the tooling cost, which includes costs of grippers, fixtures, and end-effectors.

EC is the engineering cost.

IC is the installation cost, which includes costs of set-up labor, utility hook-up and foundations.

$ACCS$ is the cost of accessories; these accessories include items such testers, part feeders, bins, and conveyors.

REC is the cost of related expenses (e.g., insurance).

TRC is the training cost.

10.8 Robot Life Cycle Cost Estimation Models

There are a number of life cycle cost models that can be used to estimate the life cycle cost of robots [14,15]. Two of such models are presented below [14,15].

10.8.1 Life Cycle Cost Model I

This model assumes that the cost of the robotic life cycle is made up of three major components: failure cost, operating cost, and initial cost. Thus, the robot life cycle cost is expressed by [1]

$$LCC_r = C_{ri} + C_{rf} + C_{ro} \tag{10.7}$$

where
  LCC_r is the cost of the robot's life cycle.
  C_{ri} is the robot's initial cost, which is made up of the cost of components such as initial spares, of training, equipment, and software.
  C_{rf} is the robot failure cost, which is made up of components such as repair cost and the cost of operational losses attributable to robot-related failures.
  C_{ro} is the robot operating cost, which is made up of elements such as salaries, cost of expendables, and preventive maintenance cost.

10.8.2 Life Cycle Cost Model II

This model assumes that the robot's life cycle cost is made up of two major components: nonrecurring cost and recurring cost. Thus, the robot's life cycle cost is defined by

$$LCC_r = C_{nr} + C_r \tag{10.8}$$

where
  LCC_r is the robot's life cycle cost.
  C_{nr} is the robot nonrecurring cost.
  C_r is the robot recurring cost.

The robot nonrecurring cost is made of elements such as listed below:

- Robot cost
- Engineering cost
- Installation cost
- Training cost
- Reliability/maintainability improvement cost

The robot's cost is made up of the cost of items such as test equipment, material handling equipment, tooling for the robot hand, and cost of robot modifications. The engineering cost is concerned with items such as system planning, layout design, programming the robot, and product redesign. The installation cost includes the cost of items such as safety equipment, vendor installation assistance, building modifications, and relocating other equipment.

The training cost is associated with items such as training of operators, programmers, and maintenance-related personnel. Finally, reliability/maintainability improvement cost is associated with the actions concerning robot-related reliability and maintainability improvement.

Some of the main elements of the recurring costs of robots are as follows:

- Operating cost
- Manpower cost
- Maintenance cost
- Inventory cost

10.9 Useful Methods for Making Financial Decisions about Robotization

There are numerous methods that can be used to make investment decisions regarding robotization [1,14]. Some of these methods are presented below [1].

10.9.1 Method I: Life Cycle Costing

This method calls for estimating the life cycle cost of the alternative projects and then comparing the end results and selecting the project with the lowest life cycle cost. The life cycle cost of a project is expressed by

$$PLC = \sum_{j=0}^{n} \left[\frac{AC_j + RC_j}{(1+i)^j} \right] - \frac{SV}{(1+i)^n} \tag{10.9}$$

where
PLC is the project life cycle cost.
n is the project life in years.
i is the annual interest rate.
RC_j is the estimated running cost in year j.
AC_j is the acquisition cost in year j.
SV is the estimated salvage value of the project under consideration.

10.9.2 Method II: Payback Period

This method is concerned with determining the length of time needed for recovering the cost of an investment. After the payback period, the robot will begin making positive returns for its entire remaining life. Thus, the robot installation payback method is expressed by [1,16]

$$N_p = \frac{I_t}{LS_t - E_{rt}}$$ (10.10)

where
N_p is the total number of years for payback of the investment cost.
I_t is the total investment for the robot and its associated accessories.
LS_t is the total annual labor savings.
E_{rt} is the total annual expense of the robot.

It is to be noted that in the payback analysis method, the robot production rate's effects can also be considered. Furthermore, in situations where costly production equipment is used, the payback is more sensitive to the production rate.

Thus, the robot payback period is defined by

$$N_{rp} = \frac{I_t}{LS_t - E_{rt} \pm \theta(LS_t + DC_{ae})}$$ (10.11)

where
N_{rp} is the total number of years for the payback on the investment in the robot.
θ is the production rate coefficient.
DC_{ae} is the annual depreciation cost of associated equipment.

EXAMPLE 10.2

Assume that the total amount of money invested for a robot and its accessories is US$100,000. The annual cost of the robot's upkeep and the expected annual labor savings are US$20,000 and US$30,000, respectively.
Calculate the payback period for the investment in the robot.
By substituting the specified data values into Equation 10.10, we get

$$N_p = \frac{100,000}{30,000 - 20,000} = 10 \text{ years}$$

Thus, the payback period for the investment in the robot is 10 years.

10.9.3 Method III: Minimum Cost Rule

In this method, costs such as capital cost and operating cost are calculated for each and every alternative, and a decision on the desirability of the investment is taken by comparing the cost totals of the alternatives. An example of the minimum cost rule method is the present value approach.

The net present value of a robot project under consideration can be estimated with the aid of the equation presented below [1,13]:

$$PV_n = \sum_{j=0}^{n} \frac{\left(PS_j - AC_j - RC_j\right)}{\left(1+i\right)^j} + \frac{V_s}{\left(1+i\right)^n} \tag{10.12}$$

where
PV_n is the net present value of the robotization project under consideration.
V_s is the expected salvage value of the robotization project under consideration.
i is the annual interest rate.
n is the robot's project life in years.
AC_j is the acquisition cost in year j.
RC_j is the expected running cost in year j.
PS_j is the expected potential savings in year j.

It is to be noted that during the alternative projects' comparing process, the maximum present value is aimed for.

10.9.4 Method IV: Return on Investment

This is probably the most celebrated method to compare alternative investments under consideration. In the case of a robotization project, once the cost of capital for investment is established, then it should be compared with the return on investment.

The return on investment is defined by [1]

$$I_r = \left(\frac{S_t}{I_{ra}}\right)(100\%) \tag{10.13}$$

where
I_r is the return on investment.
S_t is the total annual savings resulting from the robot.
I_{ra} is the total investment in the robot and its associated accessories.

10.10 Failure Data Uses with Regard to Robots and Failure Reporting and Documentation System for Robots

In various robot-related studies, failure-related data plays an important role. More specifically, some of the areas in which data become very useful are as follows [11]:

- Gaining knowledge concerning the robot's design and manufacturing-related shortcomings
- Estimating availability, reliability, and hazard rate (failure rate) of the robot
- Recommending design changes to improve the reliability of the robot or of its parts
- Supporting reliability growth-related programs for the robot
- Performing reliability cost trade-off studies for the robot
- Determining the maintenance needs of a robot and its parts
- Conducting robot design reviews
- Conducting replacement-related studies for the robot
- Performing life cycle cost-related studies for the robot

The establishment of proper failure reporting and documentation systems is essential to produce reliable and safe robots. During the designing process of such systems, it is important to pay proper attention to factors such as shown in Figure 10.3 [1].

10.11 Main Data Sources for Reliability

There are many sources for obtaining data related to product reliability. These include previous experience with similar products, tests (environmental qualification, field installation, and field demonstration), repair records, warranty claims, inspection records, and so on. In the case of robots, as they may contain electrical, mechanical, and electronic parts; the failure data for such parts may be obtained from the existing failure data banks in various industrial sectors.

An excellent document to estimate the failure rates of electronic parts is *MIL-HDBK-217E* [17]. Nonetheless, the main data sources for reliability are as follows [1,18]:

- *Laboratory tests.* In this case, under laboratory conditions, a group of similar items is tested and the failure events are recorded along with the time intervals.
- *Published literature (information).* This involves obtaining generic data for failure from open published literature (e.g., government publications, industrial reports, journals, books, and conference proceedings). It is to be noted that the data published in such documents is

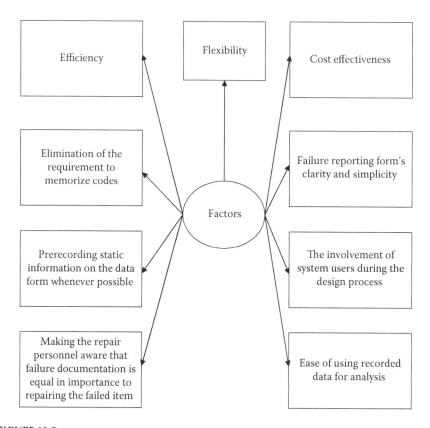

FIGURE 10.3
Factors requiring attention in designing failure reporting and documentation systems for robots.

generally in the form of failure rates, and the information on population and experience is missing; thus, it lowers the usefulness of such data.

- *Expert opinions.* In this case, data on failure is generated from the opinions expressed by experts such as equipment maintenance personnel and operators. Subsequently, their judgments are converted into quantified data with the aid of methods such as Delphi and paired comparisons.

- *Event data banks.* Specifically designed systems that are capable of accepting the details of all types of events and their occurrence times, are normally referred to as "event data banks." Data on failure and other plant management-related information is extracted by processing the available event records.

- *Field failure data collection.* This is the approach of making use of the information available in operating and maintenance records for establishing an as-complete-as-possible historical record of events. From this history, data on failure is derived.

10.12 Repair and Inspection Records-Related Requirements for Robots

Each time a robot is repaired or inspected, various types of information can be recorded. This information becomes very useful to make robot-related decisions. Robot repair records should include information on the following items at least [1,19]:

- Serial number
- Robot designation
- Name(s) of the repair person(s)
- Repair starting time and date
- Repair completion time and date
- Operating time since the previous repair
- Model number
- Repair description and the place of repair
- Type of inspection that resulted in the repair action

Similarly, the periodic inspection records should include information on the following items at least [1,19]:

- Serial number and model
- Date of inspection
- Inspector identification
- Robot designation
- Inspection location
- Inspection results
- Inspection method used
- Action to be performed for the faulty portion
- Action implemented and implementation date

Information on some possible documents, banks, and organizations to obtain data on failure for robot-related reliability studies is available in [1].

10.13 Problems

1. Describe performance parameters that play a key role in robot performance testing.
2. Describe the following robot performance testing methods:
   a. The test-robot program method
   b. The Ford method
3. What are the three broad categories of robot reliability tests? Discuss each of these categories.
4. What are the management-related tasks in a robot reliability test plan?
5. What are the robot testing and start-up safety-related factors?
6. What are the main categories of the robot project cost?
7. Assume that a robot's mean time to failure is 1600 h and its annual scheduled operating hours are 4800 h. The robot's mean time to repair is 10 h and its maintenance labor cost is US$40 per hour. Calculate the robot's annual corrective maintenance cost.
8. What are the factors that require proper attention when designing failure reporting and documentation systems for robots?
9. What are the main data sources for reliability? Discuss each of these sources.
10. List at least seven useful guidelines for robot reliability demonstration tests.

References

1. Dhillon, B.S., *Robot Reliability and Safety*, Springer-Verlag, New York, 1991.
2. Nof, S.Y., Decisions aids for planning industrial robot operations, *Proceedings of the Annual Industrial Engineering Conference*, 1982, pp. 180–185.
3. Wodzinski, M., Putting robots to the test, *Robotics Today*, 1987, pp. 17–20.
4. Robotics Industries Association, *Robotics Glossary*, Ann Arbor, Michigan, 1984.

5. Ranky, P.G., Wodzinski, M., *A Survey of Robot Test Methods with Examples, Report No. RSD-TR-1-87*, Center for Research on Integrated Manufacturing, Robot Systems Division, College of Engineering, University of Michigan, Ann Arbor, Michigan, February 1987.

6. Ranky, P.G., Ho, C.Y., *Robot Modeling: Control and Applications with Software*, Springer-Verlag, Berlin, 1985.

7. Warnecke, H.J., Test stand for industrial robots, *Proceedings of the 7th International Symposium on Industrial Robotics*, October 1977, pp. 10–15.

8. Lau, K., Hocken, R., Haight, W., *An Automatic Laser-Tracking Interferometer System for Robot Metrology, Report*, National Bureau of Standards, Gaithersburg, Maryland, March 1985.

9. Von Alven, W.H., Ed., *Reliability Engineering*, Prentice-Hall, Englewood Cliffs, New Jersey, 1964.

10. American National Standards Institute, *American National Standard for Industrial Robots and Robot Systems: Safety Requirements, ANSI/RIA R15.06, 1986*; available from the American National Standards Institute (ANSI), 1430 Broadway, New York, New York 10018, 1986.

11. Morgan, C., *Robots Planning and Implementation*, Springer-Verlag, Berlin, 1984.

12. Scott, P.B., Husband, T.M., Robotic assembly: Design, analysis and economic evaluation, *Proceedings of the 13th International Symposium on Industrial Robots*, 1983, pp. 5.12–5.29.

13. Hasegawa, Y., Evaluation and economic justification, in *Handbook of Industrial Robotics*, Ed. S.Y. Nof, John Wiley and Sons, New York, 1985, pp. 665–687.

14. Dhillon, B.S., *Life Cycle Costing for Engineers*, CRC Press, Boca Raton, Florida, 2010.

15. Dhillon, B.S., *Life Cycle Costing: Techniques, Models, and Applications*, Gordon and Breach Science Publishers, New York, 1989.

16. Miller, R.K., *Industrial Robot Handbook*, SEAI Institute, P.O. Box 590, Madison, Georgia, 1983.

17. United States Department of Defense, *MIL-HDBK-217E, Reliability Prediction of Electronic Equipment*, Washington, D.C., October 1986; available from the Naval Publications and Forms Center, 5801 Tabor Avenue, Philadelphia, Pennsylvania 19120, USA.

18. Bello, G.C., Data validation procedures, in *Reliability Data Bases*, eds. A. Amendola and A.Z. Keller, Reidel, Boston, 1987, pp. 125–132.

19. Japanese Industrial Safety and Health Association, *An Interpretation of the Technical Guidance on Safety Standards in the Use, Etc., of Industrial Robots*, 3-35-1, Shiba, Minato-Ku, Tokyo 108, Japan, 1985.

11

Mathematical Models for Analysis of Robot-Related Reliability and Safety

11.1 Introduction

Mathematical modeling is a commonly used approach to conduct various types of analysis in the area of engineering. In this case, the components of a system are represented by idealized elements assumed to have all the representative characteristics of real-life components, whose behavior is possible to be described by equations. However, it is to be noted that the degree of realism of a mathematical model depends on the type of assumptions imposed upon it.

Over the years, many mathematical models have been developed to analyze robot-related reliability and safety. Many of these models were developed using the Markov method [1,2]. Although the effectiveness of such models can vary quite considerably from one application area to another, some of them are being used quite successfully to study various types of real-life problems in industries using robot systems.

This chapter presents a number of mathematical models considered useful to perform various types of analysis of the reliability and safety of robotic systems. All of these models are based upon the Markov method.

11.2 Model I

This mathematical model represents a robot having three distinct states: working normally, failed due to a hardware or software failure, and failed due to a human error. The failed robot is repaired to revert to its working state from both the failed states. The robot state-space diagram is shown in Figure 11.1. The numerals in rectangles denote system states.

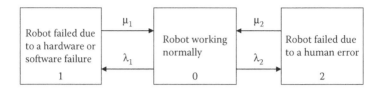

FIGURE 11.1
Robot state-space diagram.

The following four assumptions are associated with this mathematical model [1,3]:

- Robot hardware or software failure and human error rates are constant.
- The failed robot repair rates are constant.
- The repaired robot is as good as new.
- Human error and hardware or software failures are statistically independent.

The following symbols are associated with the diagram shown in Figure 11.1 and its associated equations:

$P_j(t)$ is the probability that the robot is in state j at time t; for $j = 0$ (working normally), $j = 1$ (failed due to a hardware or software failure), $j = 2$ (failed due to a human error).

λ_1 is the robot hardware or software failure rate.

λ_2 is the robot human error rate.

μ_1 is the robot repair rate from failed state 1.

μ_2 is the robot repair rate from failed state 2.

Using the Markov method, we have the following equations for the state-space diagram shown in Figure 11.1 [1,3]:

$$\frac{dP_0(t)}{dt} + (\lambda_1 + \lambda_2)P_0(t) = \mu_2 P_2(t) + \mu_1 P_1(t) \tag{11.1}$$

$$\frac{dP_1(t)}{dt} + \mu_1 P_1(t) = \lambda_1 P_0(t) \tag{11.2}$$

$$\frac{dP_2(t)}{dt} + \mu_2 P_2(t) = \lambda_2 P_0(t) \tag{11.3}$$

At time $t = 0$, $P_0(0) = 1$, $P_1(0) = 0$, and $P_2(0) = 0$.

Solving Equations 11.1 through 11.3 using Laplace transforms, we obtain

$$P_0(t) = \frac{\mu_1\mu_2}{n_1 n_2} + \left[\frac{(n_1 + \mu_1)(n_1 + \mu_2)}{n_1(n_1 - n_2)} \right] e^{n_1 t} - \left[\frac{(n_2 + \mu_1)(n_2 + \mu_2)}{n_2(n_1 - n_2)} \right] e^{n_2 t} \quad (11.4)$$

where

$$n_1, n_2 = \frac{-b \pm \left[b^2 - 4(\mu_1\mu_2 + \lambda_2\mu_1 + \lambda_1\mu_2) \right]^{1/2}}{2} \quad (11.5)$$

$$b = \lambda_1 + \lambda_2 + \mu_1 + \mu_2 \quad (11.6)$$

$$n_1 n_2 = \mu_1\mu_2 + \lambda_2\mu_1 + \lambda_1\mu_2 \quad (11.7)$$

$$n_1 + n_2 = \lambda_1 + \lambda_2 + \mu_1 + \mu_2 \quad (11.8)$$

$$P_1(t) = \frac{\lambda_1\mu_2}{n_1 n_2} + \left[\frac{\lambda_1 n_1 + \lambda_1\mu_2}{n_1(n_1 - n_2)} \right] e^{n_1 t} - \left[\frac{(\mu_2 + n_2)\lambda_1}{n_2(n_1 - n_2)} \right] e^{n_2 t} \quad (11.9)$$

$$P_2(t) = \frac{\mu_1\lambda_2}{n_1 n_2} + \left[\frac{\lambda_2 n_1 + \lambda_2\mu_1}{n_1(n_1 - n_2)} \right] e^{n_1 t} - \left[\frac{(\mu_1 + n_2)\lambda_2}{n_2(n_1 - n_2)} \right] e^{n_2 t} \quad (11.10)$$

The robot availability, $RAV(t)$, is given by

$$RAV(t) = P_0(t) \quad (11.11)$$

As time t becomes very large in Equations 11.9 through 11.11, the following steady-state probability expressions are obtained:

$$P_1 = \frac{\lambda_1\mu_2}{n_1 n_2} \quad (11.12)$$

$$P_2 = \frac{\mu_1\lambda_2}{n_1 n_2} \quad (11.13)$$

$$RAV = \frac{\mu_1\mu_2}{n_1 n_2} \quad (11.14)$$

where

P_1 is the steady-state probability of the robot being in state 1.
P_2 is the steady-state probability of the robot being in state 2.
RAV is the robot steady-state availability.

For $\mu_1 = \mu_2 = 0$, from Equations 11.4, 11.9, and 11.10, we obtain

$$P_0(t) = e^{-(\lambda_1+\lambda_2)t} \tag{11.15}$$

$$P_1(t) = \frac{\lambda_1}{(\lambda_1 + \lambda_2)}[1 - e^{-(\lambda_1+\lambda_2)t}] \tag{11.16}$$

$$P_2(t) = \frac{\lambda_2}{(\lambda_1 + \lambda_2)}[1 - e^{-(\lambda_1+\lambda_2)t}] \tag{11.17}$$

Thus, the robot reliability at time t from Equation 11.15 is

$$R_r(t) = e^{-(\lambda_1+\lambda_2)t} \tag{11.18}$$

where
$R_r(t)$ is the robot reliability at time t.

By using the formula for mean time to failure presented in Chapter 3 and Equation 11.18, we get the following expression for the mean time to robot-related failure:

$$RMTTF = \int_0^\infty e^{-(\lambda_1+\lambda_2)t} dt$$

$$= \frac{1}{\lambda_1 + \lambda_2} \tag{11.19}$$

where
$RMTTF$ is the mean time to robot failure.

EXAMPLE 11.1

Assume that a robot can fail either due to human error or other failures (i.e., hardware and software failures) and its constant human error and other failure rates are 0.002 errors per hour and 0.007 failures per hour, respectively. The robot constant repair rate from both these failure modes is 0.09 repairs per hour.

Calculate the robot steady-state availability.

By inserting the given data values into Equation 11.14, we get

$$RAV = \frac{(0.09)(0.09)}{(0.09)(0.09) + (0.002)(0.09) + (0.007)(0.09)}$$

$$= 0.9091$$

Thus, the robot steady-state availability is 0.9091.

11.3 Model II

This mathematical model represents a situation where a worker performs a time-continuous robot-related task under fluctuating environments (i.e., normal and abnormal) and he/she can commit an error in either a normal or an abnormal environment. The state-space diagram for the worker carrying out a time-continuous robot-related task is shown in Figure 11.2. The numerals in the rectangles or boxes denote system states.

This mathematical model is subjected to the following assumptions:

- Errors occur independently.
- The rate of change of the environment from normal to abnormal (or stressful) or vice versa is constant.
- Worker error rates are constant.
- The worker is executing a robot-related time-continuous task.

The following symbols are associated with the diagram shown in Figure 11.2 and its associated equations:

$P_i(t)$ is the probability of the worker being in state i at time t, for $i = 0,1,2,3$.

λ_n is the constant error rate of the worker carrying out a robot-related task in a normal environment.

λ_a is the constant error rate of the worker carrying out a root-related task in an abnormal or stressful environment.

θ_1 is the constant transition rate from a normal environment to an abnormal or stressful environment.

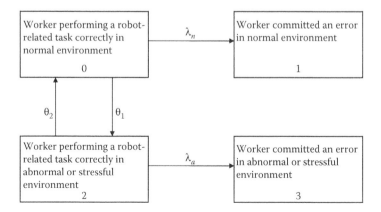

FIGURE 11.2
State-space diagram for the worker performing a robot-related task in fluctuating environments.

θ_2 is the constant transition rate from an abnormal or stressful environment to a normal environment.

Using the Markov method, we have the following set of differential equations for the state-space diagram shown in Figure 11.2 [4]:

$$\frac{dP_0(t)}{dt} + (\lambda_n + \theta_1)P_0(t) = \theta_2 P_2(t) \tag{11.20}$$

$$\frac{dP_1(t)}{dt} = \lambda_n P_0(t) \tag{11.21}$$

$$\frac{dP_2(t)}{dt} + (\lambda_a \theta_2)P_2(t) = \theta_1 P_0(t) \tag{11.22}$$

$$\frac{dP_3(t)}{dt} = \lambda_a P_2(t) \tag{11.23}$$

At time $t = 0$, $P_0(0) = 1$ and $P_1(0) = P_2(0) = P_3(0) = 0$.
Solving Equations 11.20 through 11.23, we get

$$P_0(t) = (y_1 - y_2)^{-1}[(y_2 + \lambda_a + \theta_2)e^{y_2 t} - (y_1 + \lambda_a + \theta_2)e^{x_1 t}] \tag{11.24}$$

where

$$y_1 = \frac{-c_1 + \sqrt{c_1^2 - 4c_2}}{2} \tag{11.25}$$

$$y_2 = \frac{-c_1 - \sqrt{c_1^2 - 4c_2}}{2} \tag{11.26}$$

where

$$c_1 = \lambda_n + \lambda_a + \theta_1 + \theta_2 \tag{11.27}$$

$$c_2 = \lambda_n(\lambda_a + \theta_2) + \theta_1 \lambda_a \tag{11.28}$$

$$P_1(t) = c_4 + c_5 e^{y_2 t} - c_6 e^{y_1 t} \tag{11.29}$$

where

$$c_4 = \frac{\lambda_n(\lambda_a + \theta_2)}{y_1 y_2} \tag{11.30}$$

$$c_5 = c_3(\lambda_n + c_4 y_1) \tag{11.31}$$

$$c_3 = \frac{1}{y_2 - y_1} \tag{11.32}$$

$$c_6 = c_3(\lambda_n + c_4 y_2) \tag{11.33}$$

$$P_2(t) = \theta_1 c_3(e^{y_2 t} - e^{y_1 t}) \tag{11.34}$$

$$P_3(t) = c_7[(1 + c_3)(y_1 e^{y_2 t} - y_2 e^{y_1 t})] \tag{11.35}$$

where

$$c_7 = \frac{\lambda_a \theta_1}{y_1 y_2} \tag{11.36}$$

The reliability of the worker in a fluctuating environment is given by

$$WR(t) = P_0(t) + P_1(t) \tag{11.37}$$

where
  $WR(t)$ is the reliability of the worker in carrying out a robot-related task in a fluctuating environment at time t.

By using the formula for mean time to failure presented in Chapter 3 and Equation 11.37, we obtain the following expression for the mean time to error of the worker:

$$WMTTE = \int_0^\infty WR(t)dt$$
$$= \frac{\lambda_a + \theta_1 + \theta_2}{c_2} \tag{11.38}$$

where
  $WMTTE$ is the mean time to error of the worker.

EXAMPLE 11.2

Assume that a worker is carrying out a robot-related task in a fluctuating environment (i.e., normal and abnormal [stressful]) and his/her error rates are 0.005 errors per hour and 0.009 errors per hour, respectively. The constant transition rates—from a normal to an abnormal environment and vice versa—are 0.005 per hour and 0.002 per hour, respectively. Calculate the mean time to error of the worker.

By inserting the specified data values into Equation 11.38, we get

$$WMTTE = \frac{(0.009 + 0.005 + 0.002)}{(0.005)(0.009 + 0.002) + (0.005)(0.009)}$$
$$= 160 \text{ h}$$

Thus, the mean time to error of the worker is 160 h.

11.4 Model III

This mathematical model represents a robot system made up of a robot and a safety unit; as in industry, the inclusion of safety units/systems is often practiced because of the occurrence of robot accidents involving humans. In the model, it is assumed that after the failure of the safety unit, the robot may fail with an incident or safely and the failed safety unit is repaired.

The robot system state-space diagram is shown in Figure 11.3 and the numerals in rectangles or boxes denote system states.

The model is subjected to the following four assumptions:

- The robot system fails when the robot fails.
- Failures occur independently.
- The repaired safety unit is as good as new.
- All failure and repair rates are constant.

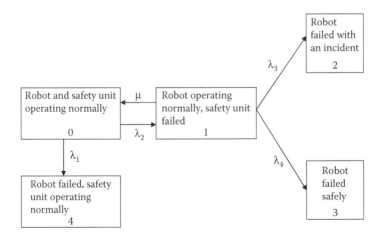

FIGURE 11.3
Robot system state-space diagram.

The following symbols are associated with the state-space diagram shown in Figure 11.3 and its associated equations:

$P_j(t)$ is the probability that the robot system is in state j at time t; for $j = 0$ (robot and safety unit operating normally), $j = 1$ (robot operating normally, safety unit failed), $j = 2$ (robot failed with an incident), $j = 3$ (robot failed safely), $j = 4$ (robot failed, safety unit operating normally).

μ is the safety unit repair rate.

λ_1 is the robot failure rate.

λ_2 is the safety unit failure rate.

λ_3 is the failure rate of the robot failing with an incident.

λ_4 is the failure rate of the robot failing safely.

Using the Markov method, we have the following set of differential equations for the state-space diagram shown in Figure 11.3 [5]:

$$\frac{dP_0(t)}{dt} + (\lambda_1 + \lambda_2)P_0(t) = \mu P_1(t) \tag{11.39}$$

$$\frac{dP_1(t)}{dt} + (\lambda_3 + \lambda_4 + \mu)P_1(t) = \lambda_2 P_0(t) \tag{11.40}$$

$$\frac{dP_2(t)}{dt} = \lambda_3 P_1(t) \tag{11.41}$$

$$\frac{dP_3(t)}{dt} = \lambda_4 P_1(t) \tag{11.42}$$

$$\frac{dP_4(t)}{dt} = \lambda_1 P_0(t) \tag{11.43}$$

At time $t = 0$, $P_0(0) = 1$ and $P_1(0) = P_2(0) = P_3(0) = P_4(0) = 0$.
Solving Equations 11.39 through 11.43 using Laplace transforms, we obtain

$$P_0(t) = e^{-Xt} + \lambda_2 \mu \left[\frac{e^{-Xt}}{(a_1 + X)(a_2 + X)} + \frac{e^{-a_1 t}}{(a_1 + X)(a_1 - a_2)} + \frac{e^{-a_2 t}}{(a_2 + X)(a_2 - a_1)} \right] \tag{11.44}$$

where

$$X = \lambda_2 + \lambda_1 \tag{11.45}$$

$$a_1, a_2 = \frac{-Y \pm (Y^2 - 4Z)^{1/2}}{2} \tag{11.46}$$

$$Y = X + \mu + \lambda_3 + \lambda_4 \tag{11.47}$$

$$Z = \lambda_3\lambda_2 + \lambda_4\lambda_2 + \lambda_3\lambda_1 + \lambda_4\lambda_1 + \mu\lambda_1 \tag{11.48}$$

$$P_1(t) = \lambda_2 \left[\frac{e^{a_1 t} - e^{a_2 t}}{a_1 - a_2} \right] \tag{11.49}$$

$$P_2(t) = \frac{\lambda_3\lambda_2}{a_1 a_2} \left[1 + \frac{a_1 e^{a_2 t} - a_2 e^{a_1 t}}{a_2 - a_1} \right] \tag{11.50}$$

$$P_3(t) = \frac{\lambda_4\lambda_2}{a_1 a_2} \left[1 + \frac{a_1 e^{a_2 t} - a_2 e^{a_1 t}}{a_2 - a_1} \right] \tag{11.51}$$

$$P_4(t) = \frac{\lambda_1}{X}(1 - e^{-Xt}) + \lambda_2\lambda_1\mu \left[\frac{1}{a_1 a_2 X} - \frac{e^{-Xt}}{X(a_1 + X)(a_2 + X)} \right.$$
$$\left. + \frac{e^{a_1 t}}{a_1(a_1 + A)(a_1 - a_2)} + \frac{e^{a_2 t}}{a_2(a_2 + X)(a_2 - a_1)} \right] \tag{11.52}$$

The robot system reliability is given by

$$RSR(t) = P_0(t) \tag{11.53}$$

where
$RSR(t)$ is the robot system reliability at time t.

By using the formula for mean time to failure presented in Chapter 3 and Equation 11.53, we get the following expression for the robot system's mean time to failure:

$$RSMTTF = \int_0^\infty RSR(t)dt$$
$$= \frac{1}{X}\left[1 + \frac{\lambda_2\mu}{Z} \right] \tag{11.54}$$

where
$RSMTTF$ is the robot system's mean time to failure.

EXAMPLE 11.3

A robot system is made up of a robot and a safety unit and the operating robot with failed safety unit can fail either safely or with an incident. Only the failed safety unit is repaired.

Calculate the robot system's mean time to failure with the aid of Equation 11.54 for the following specified data values:

$\mu = 0.004$ repairs per hour
$\lambda_2 = 0.0001$ failures per hour
$\lambda_3 = 0.0005$ failures per hour
$\lambda_4 = 0.0003$ failures per hour
$\lambda_1 = 0.0008$ failures per hour

By substituting the above given data values into Equation 11.54, we get

$$RSMTTF = \frac{1}{X}\left[1 + \frac{(0.0001)(0.004)}{Z}\right]$$

where
$X = 0.0001 + 0.0008$
$Z = (0.0005)(0.0001) + (0.0003)(0.0001) + (0.0005)(0.0008)$
$\qquad + (0.0003)(0.0008) + (0.004)(0.0008)$

$$= \frac{1}{0.0009}\left[1 + \frac{(0.0001)(0.004)}{(0.00000392)}\right]$$

$$= 1224.5\ \text{h}$$

Thus, the robot system's mean time to failure is 1224.5 h.

11.5 Model IV

This mathematical model represents a robot system having three states: operating normally, operating unsafely, and failed. The robot system is repaired from failed and unsafe operating states. The robot system state-space diagram is shown in Figure 11.4 and the numerals in circles denote the robot system states.

The model is subjected to the following assumptions:

- All robot system failure and repair rates are constant.
- The repaired robot system is as good as new.
- Failures occur independently.

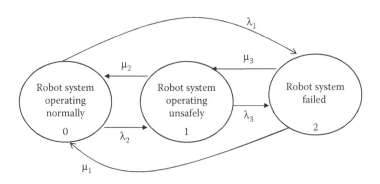

FIGURE 11.4
Robot system state-space diagram.

The following symbols are associated with the state-space diagram shown in Figure 11.4 and its associated equations:

$P_i(t)$ is the probability that the robot system is in state i at time t; for $i = 0$ (robot system operating normally), $i = 1$ (robot system operating unsafely), $i = 2$ (robot system failed).

λ_i is the robot system ith constant failure rate, where $i = 1$ means from state 0 to state 2, $i = 2$ means from state 0 to state 1, $i = 3$ means from state 1 to state 2.

μ_i is the robot system ith constant repair rate, where $i = 1$ means from state 2 to state 0, $i = 2$ means from state 1 to state 0, $i = 3$ means from state 2 to state 1.

Using the Markov method presented in Chapter 4, we have the following set of differential equations for the state-space diagram shown in Figure 11.4 [2,6,7]:

$$\frac{dP_0(t)}{dt} + (\lambda_1 + \lambda_2)P_0(t) = \mu_1 P_2(t) + \mu_2 P_1(t) \tag{11.55}$$

$$\frac{dP_1(t)}{dt} + (\mu_2 + \lambda_3)P_1(t) = \lambda_2 P_0(t) + \mu_3 P_2(t) \tag{11.56}$$

$$\frac{dP_2(t)}{dt} + (\mu_1 + \mu_3)P_2(t) = \lambda_1 P_0(t) + \lambda_3 P_1(t) \tag{11.57}$$

At time $t = 0$, $P_0(0) = 1$, and $P_1(0) = P_2(0) = 0$.
For a very large t, solving Equations 11.55 through 11.57, we get the following steady-state probability equations [7,8]:

$$P_0 = \frac{(\mu_1 + \mu_3)(\mu_2 + \lambda_3) - \lambda_3\mu_3}{D} \qquad (11.58)$$

where

$$D = (\mu_1 + \mu_3)(\mu_2 + \lambda_2 + \lambda_3) + \lambda_1(\mu_2 + \lambda_3) + \lambda_1\mu_3 + \lambda_2\lambda_3 - \lambda_3\mu_3 \qquad (11.59)$$

$$P_1 = \frac{\lambda_2(\mu_1 + \mu_3) + \lambda_1\mu_3}{D} \qquad (11.60)$$

$$P_2 = \frac{\lambda_1\lambda_3 + \lambda_1(\mu_2 + \lambda_3)}{D} \qquad (11.61)$$

where
P_0, P_1, and P_2 are the steady-state probabilities of the robot system being in states 0, 1, and 2, respectively.

By setting $\mu_1 = \mu_3 = 0$ in Equations 11.55 through 11.57 and solving the resulting equations, we get the following expression for the robot system reliability:

$$
\begin{aligned}
RSR(t) &= P_0(t) + P_1(t) \\
&= (A_1 + B_1)e^{a_1 t} + (A_2 + B_2)e^{a_2 t}
\end{aligned}
\qquad (11.62)
$$

where
$RSR(t)$ is the robot system reliability at time t.

$$a_1 = \frac{-M_1 + \sqrt{M_1^2 - 4L_2}}{2} \qquad (11.63)$$

$$a_2 = \frac{-M_1 - \sqrt{M_1^2 - 4L_2}}{2} \qquad (11.64)$$

$$M_1 = \mu_2 + \lambda_1 + \lambda_2 + \lambda_3 \qquad (11.65)$$

$$M_2 = \lambda_1\mu_2 + \lambda_1\lambda_3 + \lambda_2\lambda_3 \qquad (11.66)$$

$$A_1 = \frac{a_1 + \mu_2 + \lambda_3}{(a_1 - a_2)} \qquad (11.67)$$

$$A_2 = \frac{a_2 + \mu_2 + \lambda_3}{(a_2 - a_1)} \tag{11.68}$$

$$B_1 = \frac{\lambda_2}{(a_1 - a_2)} \tag{11.69}$$

$$B_2 = \frac{\lambda_2}{(a_2 - a_1)} \tag{11.70}$$

By integrating Equation 11.62 over the time interval $[0, \infty]$, we get the following expression for the robot system's mean time to failure:

$$RSMTTF = \int_0^\infty RSR(t)dt$$
$$= \frac{(A_1 + B_1)}{a_1} + \frac{(A_2 + B_2)}{a_2} \tag{11.71}$$

EXAMPLE 11.4

Assume that for a robot system, the following values of its associated parameters are given:

$\lambda_1 = 0.007$ failures per hour
$\lambda_2 = 0.004$ failures per hour
$\lambda_3 = 0.002$ failures per hour
$\mu_1 = 0.005$ repairs per hour
$\mu_2 = 0.002$ repairs per hour
$\mu_3 = 0.003$ repairs per hour

Calculate the steady-state probability of the robot system operating unsafely with the aid of Equation 11.60.

By inserting the given data values into Equation 11.60, we obtain

$$P_1 = \frac{(0.004)(0.005 + 0.003) + (0.007)(0.003)}{D}$$
$$= 0.460$$

where
$$D = (0.005 + 0.003)(0.002 + 0.004 + 0.002) + (0.007)(0.002 + 0.002)$$
$$+ (0.007)(0.003) + (0.004)(0.002) - (0.002)(0.003)$$

Thus, the steady-state probability of the robot system operating unsafely is 0.460.

11.6 Model V

This mathematical model represents a robot that can either fail safely or unsafely and the repair is attempted in the field; if it cannot be repaired properly, then it is taken to the repair workshop for repair [7–9]. The state-space diagram of the model is shown in Figure 11.5 and the numerals in rectangles and circles denote the robot states.

The model is subjected to the following assumptions:

- Robot failure and repair rates and the rates of taking the failed robot to the repair workshop are constant.
- Failures occur independently.
- The repaired robot is as good as new.

The following symbols are associated with the state-space diagram shown in Figure 11.5 and its associated equations:

$P_i(t)$ is the probability that the robot is in state i at time t; for $i = 0$ (robot working normally), $i = 1$ (robot failed unsafely), $i = 2$ (robot failed safely), $i = 3$ (robot in there pair workshop).

λ_1 is the robot constant unsafe failure rate.

λ_2 is the robot constant safe failure rate.

λ_3 is the constant rate of taking the safely-failed robot to the repair workshop.

λ_4 is the constant rate of taking the unsafely-failed robot to the repair workshop.

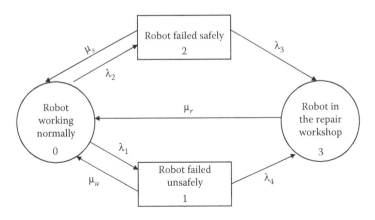

FIGURE 11.5
Robot state-space diagram.

μ_r is the constant repair rate of the robot from the repair workshop (i.e., state 3).

μ_s is the constant repair rate of the robot from the safely failed state (i.e., state 2).

μ_u is the constant repair rate of the robot from the unsafely failed state (i.e., state 1).

Using the Markov method presented in Chapter 4, we have the following set of differential equations for the state-space diagram shown in Figure 11.5 [2,6,7]:

$$\frac{dP_0(t)}{dt} + (\lambda_1 + \lambda_2)P_0(t) = \mu_u P_1(t) + \mu_s P_2(t) + \mu_r P_3(t) \tag{11.72}$$

$$\frac{dP_1(t)}{dt} + (\lambda_4 + \mu_u)P_1(t) = \lambda_1 P_0(t) \tag{11.73}$$

$$\frac{dP_2(t)}{dt} + (\lambda_3 + \mu_s)P_2(t) = \lambda_2 P_0(t) \tag{11.74}$$

$$\frac{dP_3(t)}{dt} + \mu_r P_3(t) = \lambda_3 P_2(t) + \lambda_4 P_1(t) \tag{11.75}$$

At time $t = 0$, $P_0(0) = 1$, and $P_1(0) = P_2(0) = P_3(0) = 0$.

By setting the derivatives equal to zero in Equations 11.72 through 11.75 and using the relationship $\Sigma_{i=0}^{3} P_i = 1$, we get the following steady-state probability equations [2,7]:

$$P_0 = \frac{A}{B} \tag{11.76}$$

and

$$P_i = m_i P_0, \quad \text{for } i = 1, 2, 3 \tag{11.77}$$

where

$$A = \mu_r(\lambda_4 + \mu_u)(\lambda_3 + \mu_s) \tag{11.78}$$

$$B = (\lambda_4 + \mu_u)[\mu_r(\lambda_3 + \mu_s) + \lambda_2(\mu_r + \lambda_3)] + \lambda_1(\lambda_3 + \mu_s)(\lambda_4 + \mu_r) \tag{11.79}$$

$$m_1 = \frac{\lambda_1}{\lambda_4 + \mu_u} \tag{11.80}$$

$$m_2 = \frac{\lambda_2}{\lambda_3 + \mu_s} \tag{11.81}$$

$$m_3 = \frac{\lambda_1\lambda_4(\lambda_3 + \mu_s) + \lambda_2\lambda_3(\lambda_4 + \mu_u)}{\lambda_2(\lambda_4 + \mu_u)(\lambda_3 + \mu_s)} \tag{11.82}$$

P_i is the steady-state probability of the robot being in state i, for $i = 0,1,2,3$. The robot steady-state availability and unavailability are given by

$$AV_{rs} = P_0 \tag{11.83}$$

and

$$UAV_{rs} = P_1 + P_2 + P_3 \tag{11.84}$$

where
AV_{rs} is the robot's steady-state availability.
UAV_{rs} is the robot's steady-state unavailability.

11.7 Model VI

This mathematical model is basically the same as Model V with the exception that the failed robot is not repaired. Thus, the values of μ_s, μ_u, and μ_r are equal to zero.

The model is subjected to the following two assumptions:

- Failures occur independently.
- Robot failure rates and the rates of taking the failed robot to the repair workshop are constant.

The symbols used in this mathematical model are same as for Model V.

Using the Markov method presented in Chapter 4, we have the following set of differential equations for the Figure 11.5 state-space diagram for $\mu_s = \mu_u = \mu_r = 0$ [2,6,7]

$$\frac{dP_0(t)}{dt} + (\lambda_1 + \lambda_2)P_0(t) = 0 \tag{11.85}$$

$$\frac{dP_1(t)}{dt} + \lambda_4 P_1(t) = \lambda_1 P_0(t) \tag{11.86}$$

$$\frac{dP_2(t)}{dt} + \lambda_3 P_2(t) = \lambda_2 P_0(t) \tag{11.87}$$

$$\frac{dP_3(t)}{dt} = \lambda_3 P_2(t) + \lambda_4 P_1(t) \tag{11.88}$$

At time $t = 0$, $P_0(0) = 1$, and $P_1(0) = P_2(0) = P_3(0) = 0$.
Solving Equations 11.85 through 11.88 using Laplace transforms, we obtain

$$P_0(t) = e^{-(\lambda_2 + \lambda_1)t} \tag{11.89}$$

$$P_1(t) = X\left[e^{-(\lambda_2 + \lambda_1)t} - e^{-\lambda_4 t}\right] \tag{11.90}$$

where

$$X = \frac{\lambda_1}{(\lambda_4 - \lambda_2 - \lambda_1)} \tag{11.91}$$

$$P_2(t) = Y\left[e^{-(\lambda_2 + \lambda_1 t)} - e^{-\lambda_3 t}\right] \tag{11.92}$$

where

$$Y = \frac{\lambda_2}{(\lambda_3 - \lambda_2 - \lambda_1)} \tag{11.93}$$

$$P_3(t) = 1 + Xe^{-\lambda_4 t} + Ye^{-\lambda_3 t} + Ze^{-(\lambda_2 + \lambda_1)t} \tag{11.94}$$

where

$$Z = -\frac{(X\lambda_4 + Y\lambda_3)}{(\lambda_2 + \lambda_1)} \tag{11.95}$$

The robot's reliability is given by

$$R_r(t) = P_0(t) = e^{-(\lambda_2 + \lambda_1)t} \tag{11.96}$$

where
$R_r(t)$ is the robot reliability at time t.

By integrating Equation 11.96 over the time interval $[0, \infty]$, we obtain the following expression for the robot's mean time to failure:

$$RMTTF = \int_0^\infty R_r(t)dt$$

$$= \int_0^\infty e^{-(\lambda_2+\lambda_1)t}dt$$

$$= \frac{1}{(\lambda_2 + \lambda_1)} \tag{11.97}$$

where
$RMTTF$ is the robot's mean time to failure.

EXAMPLE 11.5

A robot can fail safely or unsafely and its constant safe and unsafe failure rates are 0.008 failures per hour and 0.002 failures per hour, respectively. Calculate the robot's mean time to failure and its reliability during 8 h mission.

By substituting the specified data values into Equations 11.97 and 11.6, we get

$$RMTTF = \frac{1}{(0.008 + 0.002)}$$
$$= 100 \text{ h}$$

and

$$R_r(8) = e^{-(0.008+0.002)(8)}$$
$$= 0.9231$$

Thus, the robot's mean time to failure and reliability are 100 h and 0.9231, respectively.

11.8 Problems

1. Write an essay on mathematical models used to analyze a robot's reliability and safety.
2. Prove Equations 11.4, 11.9, and 11.10 by using Equations 11.1 through 11.3.
3. Prove Equation 11.38 by using Equation 11.37.
4. Prove that the sum of Equations 11.58, 11.60, and 11.61 is equal to unity.

5. Prove Equation 11.54 by using Equation 11.53.

6. Assume that in Figure 11.3, we have the following given values for all the five transition rates:

   a. $\lambda_1 = 0.0009$ failures per hour

   b. $\lambda_2 = 0.0002$ failures per hour

   c. $\mu = 0.003$ repairs per hour

   d. $\lambda_3 = 0.0004$ failures per hour

   e. $\lambda_4 = 0.0001$ failures per hour

   Calculate the robot system's mean time to failure.

7. Prove Equation 11.71 by using Equation 11.62.

8. Prove Equations 11.76 through 11.77 by using Equations 11.72 through 11.75.

9. Prove that the sum of Equations 11.83 through 11.84 is equal to unity.

10. Prove Equations 11.89 through 11.90, 11.92, and 11.94 with the aid of Equations 11.85 through 11.88.

References

1. Dhillon, B.S., *Applied Reliability and Quality: Fundamentals, Methods, and Procedures*, Springer, London, 2007.
2. Dhillon, B.S., *Design Reliability: Fundamentals and Applications*, CRC Press, Boca Raton, Florida, 1999.
3. Dhillon, B.S., *Robot Reliability and Safety*, Springer-Verlag, New York, 1991.
4. Dhillon, B.S., Stochastic models for predicting human reliability, *Microelectronics and Reliability*, Vol. 24, 1984, pp. 1029–1033.
5. Dhillon, B.S., Yang, N., Reliability analysis of a repairable robot system, *Journal of Quality in Maintenance Engineering*, Vol. 2, 1996, pp. 30–37.
6. Shooman, M.L., *Probabilistic Reliability: An Engineering Approach*, McGraw-Hill Book Company, New York, 1968.
7. Dhillon, B.S., *Mine Safety: A Modern Approach*, Springer, London, 2010.
8. Dhillon, B.S., Rayapati, S.N., Reliability and availability analysis of surface transit systems, *Microelectronics and Reliability*, Vol. 24, 1984, pp. 1029–1033.
9. Dhillon, B.S., *Human Reliability: With Human Factors*, Pergamon Press, New York, 1986.

Appendix A: Bibliography—Literature on the Reliability and Safety of Robot Systems

A.1 Introduction

Over the years, a large number of publications on robot-system reliability, safety, and associated areas have appeared in the form of journal articles, conference proceedings articles, technical reports, and so on. This appendix presents an extensive list of selective publications related to these topics. The period covered by the listing is 1974–2013. The main objective of this listing is to provide readers with sources for obtaining additional information on robot-system reliability, safety, and associated areas.

A.2 Publications

1. Abdallah, A., Motamed, C., Schmitt, A., Change detection for human safety in robotic environments, *Proceedings of the International Society for Optical Engineering Conference*, 1994, pp. 357–361.
2. Abdallah, A., Motamed, C., Schmitt, A., New approach for a vision-based safety device in uncontrolled robotics environments, *Proceedings of the International Society for Optical Engineering Conference*, 1995, pp. 174–181.
3. Abdi, H., Maciejewski, A.A., Nahavandi, S., Reliability maps for probabilistic guarantees of task motion for robotic manipulators, *Advanced Robotics*, Vol. 27, No. 2, 2013, pp. 81–92.
4. Abdul, S., Liu, G., Decentralized fault tolerance and fault detection of modular and reconfigurable robots with torque sensing, *Proceedings of the IEEE International Conference on Robotics and Automation*, 2008, pp. 3520–3526.
5. Abdul, S., Liu, G., Fault tolerant control of modular and reconfigurable robot with joint torque sensing, *Proceedings of the IEEE International Conference on Robotics and Biomimetics*, 2007, pp. 1236–1241.
6. Addison, J.H., Robotic safety systems and methods: Savannah River Site, *NTIS Report No. DE 8500826*, 1984; available from the National Technical Information Services (NTIS), Springfield, Virginia, USA.

7. Aghazadeh, F., Chapleski, R., Hirschfeld, R., A hazard analysis system for robotic work cells, *International Journal of Human Factors Engineering*, Vol. 8, No. 4, 1998, pp. 323–330.

8. Aghazadeh, F., Hirschfeld, R., Chapleski, R., Industrial robot use: Survey results and hazard analysis, *Proceedings of the 37th Annual Meeting of the Human Factors and Ergonomics Society*, 1993, pp. 994–998.

9. Akeel, H.A., Hardware for robotics safety systems, *Proceedings of the Interrobot West 2nd Annual Conference*, 1984, pp. 8–10.

10. Akeel, H.A., Intrinsic robot safety, *Proceedings of the Robot Safety Conference*, 1983, pp. 41–52.

11. Alayan, H., Niznik, C.A., Newcomb, R.W., Reliability of basic robot automated manufacturing networks, *Proceedings of the IEEE Southeastcon Annual Conference*, 1984, pp. 291–294.

12. Altamuro, V.M., How to achieve employee support, safety and success in your first robot installation, *Proceedings of the Robots 8 Conference*, 1984, pp. 15-1–15-8.

13. Altamuro, V.M., Working safely with iron collar worker, *National Safety News*, 1983, pp. 38–40.

14. Aluwalia, R.S., Hsu, E.Y., Sensor-based obstruction avoidance technique for a mobile robot, *Journal of Robotic Systems*, Vol. 1, 1984, pp. 331–350.

15. American Standards Institute (ANSI), *American National Standard for Industrial Robots and Robot Systems-Safety Requirements, ANSI/RIA R15.06*, New York, 1986.

16. Ancusa, V., Message redundancy in sensor networks implemented with intelligent agents, *Proceedings of the IEEE International Workshop on Robotic and Sensors Environments*, 2008, pp. 87–91.

17. Anon., Cartridge valve/brake systems increases robot safety, *Robotics World*, Vol. 1, No. 4, 1983, pp. 26–28.

18. Anon., Sensors for robot safety, *Robotics World*, Vol. 1, No. 5, 1983, pp. 16–19.

19. Anon., Protecting workers from robots, *American Machinist*, Vol. 128, No. 3, 1984, pp. 85–86.

20. Anon., Automatic robot safety shutdown system, *NASA Technical Briefs*, 1984, pp. 546–547.

21. Anon., Computing and control division colloquium on safety and reliability of complex robotic systems, *IEE Colloquium (Digest)*, No. 085, London, 1994.

22. Anon., Electrostatic liquid cleaning prevents servo-valve failures, *Robotics World*, Vol. 5, 1987, pp. 32–34.

23. Anon., HSE emphasizes a flexible approach to robot safety, *The Industrial Robot*, Vol. 14, 1987, pp. 49.

24. Anon., Robot handling provides a safe solution, *The Industrial Robot*, Vol. 12, 1985, pp. 92.

25. Anon., Robot safety: In a state of flux and jungle, *Robot News International*, 1982, pp. 3–4.

26. Anon., Safety with sophistication, *The Industrial Robot*, Vol. 11, No. 4, 1984, pp. 243–245.

27. Anon., Watchdog robot ensures safety, *Machine Design*, Vol. 55, 1983, p. 16.

28. Argote, L., Goodman, P.S., Schkade, D., The human side of robotics: How workers react to a robot, *Sloan Management Review*, 1983, pp. 33–41.

29. Asikin, D., Dolan, J., Reliability impact on planetary robotic missions, *Proceedings of the IEEE International Conference on Intelligent Robots and Systems*, 2010, pp. 4095–4100.

30. Atcitty, C.B., Robinson, D.G., Safety assessment of a robotic system handling nuclear material, *Proceedings of the ASCE Specialty Conference*, 1996, pp. 255–261.
31. Ayrulu, B., Barshan, B., Reliability measure assignment to sonar for robust target differentiation, *Pattern Recognition*, Vol. 35, No. 6, 2002, pp. 1403–1419.
32. Barabanov, V., Chirkov, A.L., Reliability of robot assisted manufacturing cells in their initial operating period, *Soviet Engineering Research*, Vol. 10, No. 1, 1990, pp. 83–85.
33. Barcheck, C., Methods for safe robotics start-up, testing, inspection and maintenance, *Proceedings of the RIA Robot Seminar*, 1985, pp. 67–73.
34. Barrett, R.J., Bell, R., Hodson, P.H., Planning for robot installation and maintenance: A safety framework, *Proceedings of the 4th British Robot Association Annual Conference*, 1981, pp. 13–22.
35. Barrett, R.J., Practical robot safety measures within a legal framework, *Proceedings of the 6th British Robot Association Annual Conference*, 1983, pp. 33–39.
36. Barrett, R.J., Robot safety and the law, in *Robot Safety*, eds. M.C. Bonney and Y.F. Yong, Springer-Verlag, Berlin, 1985, pp. 3–15.
37. Barrett, R.J., Robot safety: The HSE experience, *Proceedings of the 7th British Robot Association Annual Conference*, 1984, pp. 309–313.
38. Bary, D.J., Practical examples of safety in robot systems, *Proceedings of the International Seminar on Safety in Advanced Manufacturing*, 1987, pp. 27–36.
39. Basir, O.A., Markovian model for predicting the impact of observation conditions on the reliability of sensory systems, innovations in theory, practice and applications, *Proceedings of the IEEE International Conference on Intelligent Robots and Systems*, 1998, pp. 1646–1651.
40. Basu, A., Elnagar, A., Safety optimizing strategies for local path planning in dynamic environments, *International Journal of Robotics and Automation*, Vol. 10, No. 4, 1995, pp. 130–142.
41. Beauchamp, Y., Imbeau, D., Stobbe, T.J., Applying human factors engineering to man-robot systems, *Proceedings of the International Industrial Engineering Conference*, 1990, pp. 481–486.
42. Beauchamp, Y., Stobbe, T.J., Effects of factors on human performance in the event of an unexpected robot motion, *Journal of Safety Research*, Vol. 21, No. 3, 1990, pp. 83–96.
43. Beauchamp, Y., Stobbe, T.J., Human performance in human-robot systems, *Proceedings of the 3rd International Conference on CAD/CAM, Robotics and Factories of the Future*, 1988, pp. 33–36.
44. Becker, C. et al., Reliable navigation using landmarks, *Proceedings of the IEEE International Conference on Robotics and Automation*, 1995, pp. 401–406.
45. Ben-Lamine, M.S. et al., Mechanical impedance characteristics of robots for coexistence with humans, *Proceedings of the IEEE International Conference on Robotics and Automation*, 1997, pp. 907–912.
46. Bhatti, P.K., Rao, S.S., Reliability analysis of robot manipulators, *Proceedings of the 17th ASME Design Automation Conference*, 1987, pp. 45–53.
47. Blankenship, J., Strickland, D., Kittampton, K., Robot control for hazard prevention, *Proceedings of the American Society of Mechanical Engineers (ASME) International Computers in Engineering Conference*, 1987, pp. 164–167.
48. Bowling, A.P. et al., Reliability-based design optimization of robotic system dynamic performance, *Journal of Mechanical Design*, Vol. 129, 2007, pp. 449–454.

49. Brady, K.J. et al., Modular controller architecture for multi-arm telerobotic systems, *Proceedings of the IEEE International Conference on Robotics and Automation*, 1996, pp. 1024–1029.

50. Braman, J.M.B., Murray, R.M., Wagner, D.A., Safety verification of a fault tolerant reconfigurable autonomous goal-based robotic control system, *Proceedings of the International Conference on Intelligent Robots and Systems*, 2007, pp. 853–858.

51. Brewer, B.R., Pradhan, S., Preliminary investigation of test-retest reliability of a robotic assessment for Parkinson's disease, *Proceedings of the 32nd Annual International Conference of the IEEE EMBS*, 2010, pp. 5863–5866.

52. Broder, J., Robots seen as posing a threat to worker safety, *Los Angeles Times*, December 14, 1984.

53. Brooks, R.R., Iyengar, S.S., Robot algorithm evaluation by simulating sensor faults, *Proceedings of SPIE Conference*, Vol. 2484, 1995, pp. 394–401.

54. Brown, W.R., Ulsoy, A.G., A passive-assist design approach for improved reliability and efficiency of robots arms, *Proceedings of the IEEE International Conference on Robotics and Automation*, 2011, pp. 4927–4934.

55. Buckingham, R., Safe active robotic devices for surgery, *Proceedings of the IEEE International Conference on Systems, Man, and Cybernetics*, 1993, pp. 355–358.

56. Buckingham, R., Towards safe active robotic devices for surgery, *Industrial Robot*, Vol. 20, No. 2, 1993, pp. 8–11.

57. Burton, J., Industrial robotics: Hazards, accidents, safety applications and advanced sensor technology, *Professional Safety*, Vol. 33, 1988, pp. 28–33.

58. Caccavale, F., Walker, L.D., Observer-based fault detection for robot manipulators, *Proceedings of the IEEE International Conference on Robotics and Automation*, 1997, pp. 2881–2887.

59. Calestine, A., Park, E.H., A formal method to characterize robot reliability, *Proceedings of the Annual Reliability and Maintainability Symposium*, 1993, pp. 395–398.

60. Carlson, J., Mulphy, R.R., How UGVs physically fail in the field, *IEEE Transactions on Robotics*, Vol. 21, No. 3, 2005, pp. 423–437.

61. Carlson, J., Murphy, R.R., Nelson, A., Follow-up analysis of mobile robot failures, *Proceedings of the IEEE International Conference on Robotics and Automation*, Vol. 2004, No. 5, 2004, pp. 4987–4994.

62. Carlson, J., Murphy, R.R., Reliability analysis of mobile robots, *Proceedings of the IEEE International Conference on Robotics and Automation*, 2003, pp. 247–281.

63. Carreras, C., Walker, I.D., An interval method applied to robot reliability quantification, *Reliability Engineering and System Safety*, Vol. 70, 2000, pp. 291–303.

64. Carreras, C., Walker, I.D., Interval methods for fault-tree analysis in robotics, *IEEE Transactions on Reliability*, Vol. 50, No. 1, 2001, pp. 3–11.

65. Carreras, C., Walker, I.D., Interval methods for improved robot reliability estimation, *Proceedings of the Annual Reliability and Maintainability Symposium*, 2000, pp. 22–27.

66. Carrico, L.R., Three keys to robot safety, training, design, and implementation, *National Safety News*, 1984, pp. 81–85.

67. Cavallaro, J.R., Walker, I.D., Failure mode analysis of a proposed manipulator–based hazardous material retrieval system, *Proceedings of the ANS 7th Topical Meeting on Robotics and Remote Systems*, 1997, pp. 1096–1102.

68. Chamin, A.F., Glebov, M.R., Dytalov, V.P., Improving the reliability of robot-based sections and flexible manufacturing systems, *Chemical and Petroleum Engineering*, Vol. 24, 1988, pp. 241–245.

69. Chen, I., Effect of parallel planning on system reliability of real-time expert systems, *IEEE Transactions on Reliability*, Vol. 46, No. 1, 1997, pp. 81–87.
70. Chen, W., Hou, L., Cai, H., Research on reliability distribution method for a portable robot system, *Jiqiren/Robot*, Vol. 24, No. 1, 2002, pp. 35–38.
71. Chen, W., Hou, L., Cai, H., Two-level reliability optimizing distribution and its algorithms for series industry robot systems, *Gaojishu Tongxin/High Technology Letters*, Vol. 10, No. 2, 2000, pp. 78–81.
72. Chen, W. et al., Reliability optimizing distribution for a portable arc-welding robot system, *Harbin Gongye Daxue Xuebao/Journal of Harbin Institute of Technology*, Vol. 32, No. 1, 2000, pp. 12–14.
73. Chinellato, E. et al., Visual quality measures for characterizing planar robot grasps, *IEEE Transactions on Systems, Man, and Cybernetics*, Vol. 35, No. 1, 2005, pp. 30–41.
74. Chu, Y.X. et al., Localization algorithms performance evaluation and reliability analysis, *Proceedings of the IEEE International Conference on Robotics and Automation*, 1998, pp. 3652–3657.
75. Cohen, P.H., Chandra, M.J., Framework for the reliability modeling of robotic cells, *Proceedings of the 6th AUTOFACT Conference*, 1987, pp. 30–41.
76. Collins, J.W., Hazard prevention in automated factories, *Robotics Engineering*, Vol. 8, 1986, pp. 8–11.
77. Cop, V., Increasing the quality and reliability of industrial robots and automated workplace, in *Development in Robotics*, ed. B. Rooks, North-Holland, Amsterdam, 1983, pp. 89–98.
78. Corbel, D., Company, O., Pierrot, F., From a 3-DOF parallel redundant archi robot to a auto-calibrated archi robot, *Proceedings of the ASME International Design Engineering Technical Conference and Computers and Information in Engineering Conference*, 2007, pp. 847–855.
79. Crane, C.D. et al., Faster than real time robot simulation for plan development and robot safety, *Proceedings of the Conference on Remote Systems Technology*, 1996, pp. 14–16.
80. Crowder, R.M. et al., Maintenance of robotic systems using hypermedia and case-based reasoning, *Proceedings of the IEEE International Conference on Robotics and Automation*, 2000, pp. 2422–2427.
81. Dashiu, G. et al., Robotic assembly operation reliability, *Proceedings of the ASME Conference on Flexible Assembly Systems*, 1990, pp. 65–70.
82. Davies, B.L., Hibberd, R., Safe communication system for wheelchair-mounted medical robots, *Computing and Control Engineering Journal*, Vol. 6, No. 5, 1995, pp. 26–221.
83. Davies, B.L., Ng, W.S., Hibberd, P.D., Prostatic resection: An example safe robotic surgery, *Robotica*, Vol. 11, 1993, pp. 561–566.
84. Desai, M. et al., Effects of changing reliability on trust of robot systems, *Proceedings of the IEEE International Conference on Human-Robot Interaction*, 2012, pp. 73–80.
85. Devianayagam, S., Human factors concerns in industrial robot applications, *Proceedings of the Human Factors Society Annual Conference*, 1984, pp. 40–44.
86. Dhillon, B.S., Aleem, M.A., Report on robot reliability and safety in Canada: A survey of robot users, *Journal of Quality in Maintenance Engineering*, Vol. 6, No. 1, 2000, pp. 61–74.
87. Dhillon, B.S., Anude, O.C., Robot safety and reliability: A review, *Microelectronics and Reliability*, Vol. 33, No. 3, 1993, pp. 413–429.

88. Dhillon, B.S., Fashandi, A.R.M., Liu, K.L., Robot systems reliability and safety: A review, *Journal of Quality in Maintenance Engineering*, Vol. 8, No. 3, 2002, pp. 170–212.

89. Dhillon, B.S., Fashandi, A.R.M., Robotic systems probabilistic analysis, *Microelectronics and Reliability*, Vol. 37, No. 2, 1997, pp. 211–224.

90. Dhillon, B.S., Fashandi, A.R.M., Safety and reliability assessment techniques in robotics, *Robotica*, Vol. 15, No. 6, 1997, pp. 701–708.

91. Dhillon, B.S., Fashandi, A.R.M., Stochastic analysis of a robot machine with duplicate safety units, *Journal of Quality in Maintenance Engineering*, Vol. 5, No. 2, 1999, pp. 114–127.

92. Dhillon, B.S., Li, Z., Stochastic analysis of a maintainable robot-safety system with common-cause failures, *Journal of Quality in Maintenance Engineering*, Vol. 10, No. 2, 2004, pp. 136–147.

93. Dhillon, B.S., On robot reliability and safety: Bibliography, *Microelectronics and Reliability*, Vol. 27, 1987, pp. 105–118.

94. Dhillon, B.S., *Robot Reliability and Safety*, Springer-Verlag, New York, 1991.

95. Dhillon, B.S., Yang, N., Availability analysis of a robot with safety system, *Microelectronics and Reliability*, Vol. 36, No. 2, 1996, pp. 169–177.

96. Dhillon, B.S., Yang, N., Formulas for analyzing a redundant robot configuration with a built-in system, *Microelectronics and Reliability*, Vol. 37, No. 4, 1997, pp. 557–563.

97. Dhillon, B.S., Yang, N., Reliability analysis of a repairable robot system, *Journal of Quality in Maintenance Engineering*, Vol. 2, No. 2, 1996, pp. 30–37.

98. Dixon, W.E. et al., Fault detection for robot manipulators with parametric uncertainty: A prediction-error-based approach, *IEEE Transactions on Robotics and Automation*, Vol. 16, No. 6, 2000, pp. 689–699.

99. Dixon, W.E. et al., Fault detection for robot manipulators with parametric uncertainty: A prediction error-based approach, *Proceedings of the IEEE International Conference on Robotics and Automation*, Vol. 4, 2000, pp. 3628–3634.

100. Doming, E.E. et al., Robotic and nuclear safety for an automated/tele-operated glove box system, *Proceedings of the Conference on Remote Systems Technology*, 1992, pp. 21–25.

101. Dragone, W. et al., Failsafe brakes keep robot arms in position, *Robotics World*, Vol. 2, 1984, pp. 20–22.

102. Duggan, R.H., Jones, R.H., Khodabandehloo, K., Towards developing reliability and safety related standards using systematic methodologies, robot technology and applications, *Proceedings of the 1st Robotics Europe Conference*, 1984, pp. 90–107.

103. Eberts, R., Salvendy, G., The contributions of cognitive engineering to the safe design and operation of CAM and robotics, *Journal of Occupational Accidents*, Vol. 8, 1986, pp. 49–67.

104. Ehrenweber, R., Testing quality characteristics with a robot, *Kunststoffe Plast Europe*, Vol. 86, No. 4, 1996, pp. 12–13.

105. Etherton, J.R., Automated maintainability records and robot safety, *Proceedings of the Annual Reliability and Maintainability Symposium*, 1987, pp. 135–140.

106. Etherton, J.R., Safe maintenance guidelines for robotic workstation, Report No. 88-108, National Institute for Occupational Safety and Health (NIOSH), Morgantown, West Virginia, 1988.

107. Etherton, J.R., Systems considerations on robot and effector speed as a risk factor during robot maintenance, *Proceedings of the 8th International Safety Conference*, 1987, pp. 343–347.

108. Faber, F., Bennewitz, M., Behnke, S., Controlling the gaze direction of a humanoid robot with redundant joints, *Proceedings of the 17th IEEE International Symposium on Robot and Human Interactive Communication*, 2008, pp. 413–418.

109. Falotico, E. et al., Heat stabilization based on a feedback error learning in a humanoid robot, *Proceedings of the IEEE International Symposium on Robot and Human Interactive Communication*, 2012, pp. 449–454.

110. Fidan, B., Anderson, B.D.O., Switching control for robust autonomous robot and vehicle platoon formation maintenance, *Proceedings of the Mediterranean Conference on Control and Automation*, 2007, pp. 1–6.

111. Fluehr, P., Industrial robots and safety report, *Werkstatt and Betrieb*, Vol. 120, No. 9, 1987, pp. 733–734.

112. Folio, D., Cadenat, V., A Redundancy-based scheme to perform safe vision–based tasks amidst obstacles, *Proceedings of the IEEE International Conference on Robotics and Biometrics*, 2006, pp. 13–18.

113. Fox, D., Taking a good, hard look at robotic safety, *Robotics World*, Vol. 17, No. 1, 1999, pp. 26–29.

114. Friedrich, W.E., Robotic handling: Sensors increase reliability, *Industrial Robot*, Vol. 22, No. 4, 1995, pp. 23–26.

115. Fryman, J., New robot safety standard: New features in the design and installation of industrial robots, *Robotics World*, Vol. 18, No. 2, 2000, p. 40.

116. Gainer, C.A., Jiang, B.C., A cause and effect analysis of industrial robot accidents from four countries, *Proceedings of the Robots 11th and 17th International Symposium on Industrial Robots*, 1987, pp. 9.1–9.9.

117. Gao, X., Liao, M., Wu, X., Run-time error detection of space-robot based on adaptive redundancy, *Aircraft Engineering and Aerospace Technology: An International Journal*, Vol. 81, No. 1, 2009, pp. 14–18.

118. Garcia, M.G.A. et al., Modeling a fault-tolerant multi agent system for the control of a mobile robot using MASE methodology, *Proceedings of the 5th WSEAS International conference on Applied Computer Science*, 2006, pp. 736–744.

119. Gertman, D.I., Bruemmer, D.J., Hartley, R.S., Improving emergency response and human–robotic performance, *Proceedings of the IEEE 8th Human Factors and Power Plants and HPRCT 13th Meeting*, 2007, pp. 334–340.

120. Ghosh, K., Lemay, C., Man–machine interactions in robotics and their effect on the safety of the work place, *Proceedings of the Robot 9th Annual Conference*, 1985, pp. 19.1–19.8.

121. Ghosh, K., Safety considerations in robotics installations, *Proceedings of the 3rd Canadian CAM/CAM and Robotics Conference*, 1984, pp. 7.9–7.15.

122. Ghosh, K., Sevingy, A., Myers, M.L., Safety problems in robotics systems and some methods for their solution, *Proceedings of the 3rd International Conference on Human Factors in Manufacturing*, 1984, pp. 724–732.

123. Ghoshray, S., Yen, K.K., Comprehensive robot collision avoidance scheme by two-dimensional geometric modeling, *Proceedings of the IEEE International Conference on Robotics and Automation*, 1986, pp. 1087–1092.

124. Gopinath, P. et al., Representation and execution support for reliable robot applications, *Proceedings of the Annual Symposium on Reliability in Distributed Systems*, 1990, pp. 96–103.

125. Gordon, J.W., Curry, J.J., Reliability analysis of the RRV-1 robot, *Proceedings of the 33rd Conference on Remote Systems Technology*, 1987, pp. 210–214.

126. Graham, J.H., Fuzzy logic approach for safety and collision avoidance in robotic systems, *International Journal of Human Factors in Manufacturing*, Vol. 5, No. 4, 1995, pp. 447–457.

127. Graham, J.H., Meagher, J.F., A sensory-based robotic safety system, *IEEE Proceedings*, Vol. 132, No. 4, 1985, pp. 183–189.

128. Graham, J.H., Meagher, J.F., Derby, S.J., A safety and collision avoidance system for industrial robots, *IEEE Transactions on Industry Applications*, Vol. 22, 1986, pp. 195–203.

129. Graham, J.H., Research issues in robot safety, *Proceedings of the IEEE International Conference on Robotics and Automation*, 1988, pp. 1854–1855.

130. Granda, T. et al., Evolutionary role of humans in the human–robot system, *Proceedings of the Human Factors Society 34th Annual Meeting*, 1990, pp. 664–668.

131. Griffin, K.G., Safety requirements for industrial robotic applications, *Proceedings of the 1st International Robot Conference*, 1986, pp. 220–225.

132. Gu, Y., Zhou, H., Ma, H., Group-control on multiped robot and method of reliability, *Jiqiren/Robot*, Vol. 24, No. 2, 2002, pp. 140–144.

133. Guiochet, J., Tondu, B., Baron, C., Integration of UML in human factors analysis for safety of a medical robot for tele-echography, *Proceedings of the IEEE International Conference on Intelligent Robots and Systems*, Vol. 4, 2003, pp. 3212–3217.

134. Haass, U.L., Kuntze, H.B., Schill, W., A surveillance system for obstacle recognition and collision avoidance control in robot environment, *Proceedings of the 2nd International Conference on Robot Vision and Sensory Controls*, 1982, pp. 357–366.

135. Halberstam, E. et al., A robot supervision architecture for safe and efficient space exploration and operation, *Proceedings of the 10th Biennial International Conference on Engineering, Construction, and Operations in Challenging Environments*, 2006, pp. 1–8.

136. Hamilton, D.I., Cavallaro, J.R., Walker, I.D., Risk and fault tolerance analysis for robotics and manufacturing, *Proceedings of the 8th Mediterranean Electrotechnical Conference*, 1996, pp. 250–255.

137. Hamilton, D.L., Walker, I.D., Bennett, J.K., Fault tolerance versus performance metrics for robot systems, *Reliability Engineering & System Safety*, Vol. 53, No. 3, 1996, pp. 309–318.

138. Hamilton, J.E., Hancock, P.A., Robotic safety: Exclusion guarding for industrial operations, *Journal of Occupational Accidents*, Vol. 8, No. 1–2, 1986, pp. 69–78.

139. Harless, M., Donath, M., An intelligent safety system for unstructured human–robot interaction, *Proceedings of the Robots 9th Conference and Exposition*, 1985, pp. 117–120.

140. Harpel, B.M. et al., Analysis of robots for hazardous environments, *Proceedings of the Annual Reliability and Maintainability Symposium*, 1997, pp. 111–116.

141. Hartmann, G., Safety features illustrated in the use of industrial robots employed in production in precision mechanical/electrical industries and manufacture of appliances, *Journal of Occupational Accidents*, Vol. 8, 1986, pp. 91–98.

142. Hasegawa, Y., Industrial robot application model design for labor saving and safety promotion in press operations, *Proceedings of the 4th International Symposium on Industrial Robots*, 1974, pp. 210–213.

143. Hasegawa, Y., Sugimoto, N., Industrial safety and robots, *Proceedings of the 12th International Symposium on Industrial Robots*, 1982, pp. 9–15.

144. Haugan, K.M., Reliability in industrial robots for spray-gun applications, *Proceedings of the 2nd Conference on Industrial Robot Technology*, 1974, pp. E.7.93–E.7.98.
145. Hazon, N., Kaminka, G.A., On redundancy, efficiency, and robustness in coverage for multiple robots, *Robotics and Autonomous Systems*, Vol. 56, 2008, pp. 1102–1114.
146. Heemskerk, C., Bosman, R., HERA: A reliable and safe space robot, *Proceedings of the IEEE International Conference on Robotics and Automation*, 1993, pp. 2799–2801.
147. Helander, M., Ergonomics and safety considerations in the design of robotics workplace: A review and some priorities for research, *International Journal of Industrial Ergonomics*, Vol. 6, No. 2, 1990, pp. 127–149.
148. Helander, M., Karawan, M.H., Methods for field evaluation of safety in a robotics workplace, in *Ergonomics of Hybrid Automated System I*, ed. W. Karwowski, Elsevier, Amsterdam, 1988, pp. 403–410.
149. Helander, M., Karawn, M.H., Etherton, J., A model of human reaction time to dangerous robot arm movements, *Proceedings of the Human Factors Society 31st Annual Meeting*, 1987, pp. 191–195.
150. Henkel, S., Robots and safety: An industry overview, *Robotics Age, The Journal of Intelligent Machine*, Vol. 7, 1985, pp. 26–28.
151. Henstridge, F., Efficiency and safety found in robotic surveying, *Public Works*, Vol. 125, No. 2, 1994, pp. 26–28.
152. Hirschfeld, RA., Aghazadeh, F., Chapleski, R.C., Survey of robot safety in industry, *International Journal of Human Factors in Manufacturing*, Vol. 3, No. 4, 1993, pp. 369–379.
153. Hishino, S., Ota, J., Design of an automated transportation system in a seaport container terminal for the reliability of operating robots, *Proceedings of the International Conference on Intelligent Robots and Systems*, 2007, pp. 4259–4264.
154. Hosaka, S., Shimizu, Y., Hayashi, T., Development of a functional fail-safe control for advanced robots, *Advanced Robotics*, Vol. 8, No. 5, 1994, pp. 477–495.
155. Hoshino, S., Ota, J., Reactive robot control with hybrid operational models in a seaport container terminal considering system reliability, *Proceedings of the IEEE/RSJ International Conference on Intelligent Robots and Systems*, 2008, pp. 143–148.
156. Hudson, M.P., Pull, D.V., Intangible light barriers for the safeguarding of industrial robot cells and other automated high risk work areas, *Proceedings of the 8th British Robot Association Annual Conference*, 1985, pp. 177–179.
157. Hulfachor, R., Safety considerations and robotic welding, *Robotics Today*, 1987, pp. 24–26.
158. Husband, T.M., Managing robot maintenance, *Proceedings of the 6th British Robot Association Annual Conference*, 1983, pp. 53–60.
159. Ikuta, K., Nokata, M., General evaluation method of safety for human-care robots, *Proceedings of the IEEE International Conference on Robotics and Automation*, 1999, pp. 2065–2072.
160. Inagaki, S., Sato, K., Summary report on the status of safety engineering for industrial robots in the United States, *Katakana/Robot*, Vol. 114, 1997, pp. 17–21.
161. Isozaki, Y., Ohnishi, K., Nagashima, T., New generation maintenance and inspection robots for nuclear power plant, *Proceedings of the International Conference on Offshore Mechanics and Arctic Engineering*, 1997, pp. 77–84.

162. Jacob, P., Surface ESD (ESDFOS) in assembly fab machineries as a functional and reliability risk-failure analysis, tool diagnosis and on-site-remedies, *Microelectronics Reliability*, Vol. 48, 2008, pp. 1608–1612.
163. Jiang, B.C. et al., Evaluation machine guarding technique for robot guarding, *Robotics and Autonomous Systems*, Vol. 7, No. 4, 1991, pp. 299–308.
164. Jiang, B.C., Gainer, C.A., A cause and effect analysis of robot accidents, *Journal of Occupational Accidents*, Vol. 9, 1987, pp. 27–45.
165. Jiang, B.C., Otto, S.H., Procedure analysis for robot system safety, *International Journal of Industrial Ergonomics*, Vol. 6, No. 2, 1990, pp. 96–117.
166. Jin, Q., Sugasawa, Y., Seya, K., Probabilistic behavior and reliability analysis for a multi-robot system by applying Petri Net and Markov renewal process theory, *Microelectronics and Reliability*, Vol. 29, 1989, pp. 993–1001.
167. Jones, R.H., Dawson, S.J., People and robots: Their safety and reliability, *Proceedings of the 7th British Robot Association Conference*, 1983, pp. 156–158.
168. Jones, R.H., Dawson, S.J., Strategies for ensuring safety with industrial robot systems, *Omega International Journal of Management Science*, Vol. 14, 1986, pp. 287–297.
169. Jones, R.H., Dawson, S.J., The role of hardware, software and people in safe-guarding robot production systems, *Proceedings of the 15th International Symposium on Industrial Robots*, 1985, pp. 557–568.
170. Junyao, G. et al., Fault-tolerant and high reliability space robot design and research, *Proceedings of the International Joint Conference on Neural Networks*, 2008, pp. 2413–2417.
171. Kabayashi, F., Arai, F., Fakuda, T., Sensor selection by reliability based on possibility measure, *Proceedings of the IEEE International Conference on Robotics and Automation*, 1999, pp. 2614–2619.
172. Kanatani, K., Ohta, N., Optimal robot self-localization and reliability evaluation, *Lecture Notes in Computer Science*, Vol. 1407, 1998, pp. 796–800.
173. Kanda, T., Ishiguro, H., An approach for a social robot to understand human relationships: Friendship estimation through interaction with robots, *Interaction Studies*, Vol. 7. No. 3, 2006, pp. 369–403.
174. Kanda, T., Ishiguro, H., Friendship estimation model for social robots to understand human relationships, *Proceedings of the IEEE International Workshop on Robot and Human Interactive Communication*, 2004, pp. 539–544.
175. Kannan, B., Parkar, L.E., Metrics for quantifying system performance in intelligent, fault-tolerant multi-robot teams, *Proceedings of the IEEE International Conference on Intelligent Robots and Systems*, 2007, pp. 951–958.
176. Karwowski, W. et al., Estimation of safe distance from the robot arm as a guide for limiting slow speed of robot motions, *Proceedings of the Human Factors Society Annual Conference*, 1993, pp. 992–996.
177. Karwowski, W. et al., Human perception of the work envelope of an industrial robot, *Journal of Occupational Accidents*, Vol. 10, 1988, pp. 116–120.
178. Karwowski, W. et al., Perception of safety zone around industrial robot, *Journal of Occupational Accidents*, Vol. 10, 1988, pp. 230–235.
170. Karwowski, W., Rahimi, M., Worker selection of safe speed and idle condition in simulated monitoring of two industrial robots, *Ergonomics*, Vol. 34, No. 5, 1991, pp. 531–546.
180. Kehoe, E.J., Practical robot safety, *Robotics Today*, 1985, pp. 38–41.

181. Kemeny, Z., Design and evaluation environment for collision-free motion planning of cooperating redundant robots, *Periodica Polytechnica Ser. El, Eng.*, Vol. 43, No. 3, 1999, pp. 189–198.

182. Khodabandehloo, K., Analyses of robot systems using fault and event trees: Case studies, *Reliability Engineering & System Safety*, Vol. 53, No. 3, 1996, pp. 247–264.

183. Khodabandehloo, K., Duggan, F., Husband, T.M., Reliability assessment of industrial robots, *Proceedings of the 14th International Symposium on Industrial Robots*, 1984, pp. 203–220.

184. Khodabandehloo, K., Duggan, F., Husband, T.M., Reliability of industrial robots: A safety viewpoint, *Proceedings of the 7th British Robot Association Conference*, 1984, pp. 143–147.

185. Khodabandehloo, K., Robot safety and reliability, in *Human-Robot Interaction*, eds. M. Rahimi and W. Karwowski, Taylor & Francis, London, 1992, pp. 121–160.

186. Kim, S., Hamel, W.R., Fault detection of tool/load grasping for telerobotics using neural networks, *Proceedings of the International Conference on Advanced Robotics*, 2005, pp. 864–869.

187. Kim., S. et al., Error analysis and calibration of camera/LRF sensor head for telerobotic systems, *Proceedings of the International Conference on Intelligent Robots and Systems*, 2006, pp. 5195–5200.

188. Kinsley, J., Sophisticated sensor system enhances robot safety, *Electrical Construction and Maintenance*, Vol. 83, 1984, pp. 53–54.

189. Kittiampton, K., Sneckenberger, J., A safety control system for a robotic work station, *Proceedings of the American Control Conference*, 1985, pp. 90–93.

190. Kobayashi, F., Arai, F., Fukuda, T., Sensor selection by reliability based on possibility measure, *Proceedings of the IEEE International Conference on Robotics and Automation*, Vol. 4, 1999, pp. 2614–2619.

191. Kochekali, H. et al., Factor affecting robot performance, *Industrial Robot*, Vol. 18, No. 1, 1991, pp. 9–13.

192. Koenig, S., Simmons, R.G., Unsupervised learning of probabilistic models for robot navigation, *Proceedings of the IEEE International Conference on Robotics and Automation*, 1996, pp. 2301–2308.

193. Koker, R., Reliability-based approach to the inverse kinematics solution of robots using Elman's networks, *Engineering Applications of Artificial Intelligence*, Vol. 18, No. 6, 2005, pp. 685–693.

194. Korayem, M.H., Iravani, A., Improvement of 3P and 6R mechanical robots reliability and quality applying FMEA and QFD approaches, *Robotics and Computer-Integrated Manufacturing*, Vol. 24, 2008, pp. 472–487.

195. Kotake, S., Safety education for people working with industrial robots, *Advanced Robotics*, Vol. 3, No. 1, 1989, pp. 75–80.

196. Kuivanen, R., Experiences from the use of an intelligent safety sensor with industrial robots, in *Ergonomics of Hybrid Automated System I*, ed. W. Karwowski, Elsevier, Amsterdam, 1988, pp. 553–559.

197. Kuntze, H.B., Schill, W., Methods for collision avoidance in computer controlled industrial robots, *Proceedings of the 12th International Symposium on Industrial Robots*, 1982, pp. 519–530.

198. Lankenau, A., Meyer, O., Krieg-Brueckner, B., Safety in robotics: The Bremen autonomous wheelchair, *Proceedings of the International Workshop on Advanced Motion Control*, 1998, pp. 524–529.

199. Lauridsen, K., Reliability of remote manipulator systems for use in radiation environments, *Proceedings of the Computing and Control Division Colloquium on Safety and Reliability of Complex Robotic Systems*, 1994, pp. 1.1–1.5.

200. Lee, C., Xu, Y., Online, interactive learning of gestures for human/robot interfaces, *Proceedings of the IEEE International Conference on Robotics and Automation*, 1996, pp. 2982–2987.

201. Lee, S. et al., Human and robot integrated teleoperation, *Proceedings of the IEEE International Conference on Systems, Man, and Cybernetics*, Vol. 2, 1998, pp. 1213–1218.

202. Lendvay, M., Accelerating reliability test of electromechanical contacts to robot controlling, *Proceedings of the International Conference on Intelligent Engineering Systems*, 1997, pp. 421–425.

203. Lester, W.A., Lannon, R.P., Bellandi, R., Recommendations for maintenance of robots, Part I, *Industrial Engineering*, Vol. 17, 1985, pp. 28–30.

204. Lester, W.A., Lannon, R.P., Bellandi, R., Recommendations for maintenance of robots, Part II, *Industrial Engineering*, Vol. 17, 1985, pp. 30–35.

205. Letchmanan, R. et al., Fault evaluation of relative-coupled BLDC drives for multi-facet mobile robot with distributed speed factors, *Proceedings of the IEEE Conference on Vehicle Power and Propulsion*, 2006, pp. 1–6.

206. Leuschen, M.L., Walker, I.D., Cavallaro, J.R., Evaluating the reliability of prototype degradable systems, *Reliability Engineering and System Safety*, Vol. 72, No. 1, 2001, pp. 9–20.

207. Leuschen, M.L., Walker, I.D., Cavallaro, J.R., Robot reliability through fuzzy Markov models, *Proceedings of the Annual Reliability and Maintainability Symposium*, 1998, pp. 209–214.

208. Lewis, C.L., Maciejewski, A.A., Example of failure tolerant operation of a kinematically redundant manipulator, *Proceedings of the IEEE International Conference on Robotics and Automation*, 1994, pp. 1380–1387.

209. Lim, H., Tanie, K., Human safety mechanisms of human-friendly robots: Passive viscoelastic trunk and passively movable base, *International Journal of Robotics Research*, Vol. 19, No. 4, 2000, pp. 307–335.

210. Lin, C., Wang, M.J., Hybrid fault tree analysis using fuzzy sets, *Reliability Engineering & System Safety*, Vol. 58, No. 3, 1997, pp. 205–213.

211. Linger, M., How to design safety systems for human protection in robot applications, *Proceedings of the 14th International Symposium on Industrial Robots*, 1984, pp. 119–129.

212. Linger, M., Sjostorm, H., Palmers, G., How to design safety functions in the control system and for the grippers of industrial robot, *Proceedings of the 15th International Symposium on Industrial Robots*, 1985, pp. 569–577.

213. Liu, B. et al., Detection of range errors due to occlusion in separated transceiver LADARS, *Proceedings of the ICARCV Conference on Control, Automation, Robotics, and Vision*, 2004, pp. 443–448.

214. Liu, T.S., Wang, J.D., Reliability approach to evaluating robot accuracy performance, *Mechanism & Machine Theory*, Vol. 29, No. 1, 1994, pp. 83–94.

215. Luck, R., Ray, A., Redundancy management and failure detection of ultrasonic ranging sensors for robotic applications, *Proceedings of the USA–Japan Symposium on Flexible Automation*, 1988, pp. 557–564.

216. Lueth, T., Extensive manipulation capabilities and reliable behavior at autonomous robot assembly, *Proceedings of the IEEE International Conference on Robotics and Automation*, 1994, pp. 3495–3500.

217. Lueth, T.C., Nassal, U.M., Rembold, U., Reliability and integrated capabilities of locomotion and manipulation for autonomous robot assembly, *Robotics and Autonomous Systems*, Vol. 14, No. 2–3, 1995, pp. 185–198.

218. Lueth, T.C., Rembold, U., Extensive manipulation capabilities and reliable behavior at autonomous robot assembly, *Proceedings of the IEEE International Conference on Robotics and Automation*, 1994, pp. 3495–3500.

219. Lumelsky, V., Cheung, E., Towards safe real-time robot teleoperation: Automatic whole sensitive arm collision avoidance frees the operator for global control, *Proceedings of the IEEE International Conference on Robotics and Automation*, 1991, pp. 797–802.

220. Luridsen, K., Christensen, P., Kongso, H.E., Assessment of the reliability of robotic system for use in radiation environments, *Reliability Engineering and System Safety*, Vol. 53, 1996, pp. 265–276.

221. Luyi, C., Xiaoqing, G., Zhenyuan, X., Reliability forecasting for industrial robot based on genetic algorithm and RBF neural network, *Proceedings of the 2nd International Conference on Mechanical and Electronics Engineering*, 2010, pp. V2-362–V2-365.

222. Lynne, E.P., Alliance: An architecture for fault tolerant multi-robot operation, *IEEE Transactions on Robotics and Automation*, Vol. 14, No. 2, 1998, pp. 220–240.

223. Maier, T., Volta, G., Wilikens, M., Reliability of robotics: An overview with identification of specific aspects related to remote handling in fusion machines, *Fusion Engineering and Design*, Vol. 29, Part C, 1995, pp. 286–297.

224. Marton, T., Pulasi, J.L., Assessment and development of HF related safety design for industrial robots, *Proceedings of the Human Factors Society 31st Annual Meeting*, 1987, pp. 176–180.

225. Matsuura, D., Iwatsuki, N., Okada, M., Redundancy optimization of hyper redundant robots based on movability and assistability, *Proceedings of the International Conference on Intelligent Robots and Systems*, 2007, pp. 2534–2539.

226. McAlinden, J.J., Using robotics as occupational health and safety control strategy, *The Industrial Robot*, Vol. 22, No. 1, 1995, pp. 14–17.

227. McCulloch, W.H., Safety analysis requirements for robotic systems in DOE nuclear facilities, *Proceedings of the ASCE Specialty Conference*, 1996, pp. 235–240.

228. Mcinroy, J.E., Saridis, G.N., Entropy searches for robotic reliability assessment, *Proceedings of the IEEE International Conference on Robotics and Automation*, 1993, pp. 935–940.

229. McInroy, J.E., Saridis, G.N., Reliable automatic plan selection for visual robotic positioning, *Proceedings of the 30th Conference on Decision and Control*, 1991, pp. 1567–1572.

230. McKinnon, R., Robots: Are they automatically safe?, *Protection*, Vol. 17, No. 5, 1980, pp. 5–6.

231. Meagher, J., Derby, S., Graliam, J., Robot safety/collision avoidance, *Professional Safety*, Vol. 28, No. 12, 1983, pp. 14–16.

232. Menon, C., Murphy, M., Sitti, M., Gecko inspired surface climbing robots, *Proceedings of the IEEE Conference on Robotics and Automation*, 2004, pp. 431–436.

233. Michaelson, D.G., Jiang, J., Modelling of redundancy in multiple mobile robots, *Proceedings of the American Control Conference*, 2000, pp. 1083–1087.

234. Mihalasky, J., The impact of robots on product reliability, *Proceedings of the Annual Reliability and Maintainability Symposium*, 1985, pp. 464–466.

235. Mladnov, M., Mock, M., Grosspietsch, K.E., Fault monitoring and correction in a walking robot using LMS filters, *Proceedings of the International Workshop on Intelligent Solutions in Embedded Systems*, 2008, pp. 1–10.
236. Monteverde, V., Tosunoglu, S., Development and application of a fault tolerance measure for serial and parallel robotic structures, *International Journal of Modeling and Simulation*, Vol. 19, No. 1, 1999, pp. 45–51.
237. Monteverde, V., Tosunoglu, S., Effect of kinematic structure and dual actuation on fault tolerance of robot manipulators, *Proceedings of the IEEE International Conference on Robotics and Automation*, 1997, pp. 2902–2907.
238. Moon, I., Joung, S. Kum, Y., Safe and reliable intelligent wheelchair robot with human-robot interaction, *Proceedings of the IEEE International Conference on Robotics and Automation*, Vol. 4, 2002, pp. 3595–3600.
239. Moorign, B., Pack, T., Aspect of robot reliability, *Robotica*, Vol. 5, Part 3, 1987, pp. 232–230.
240. Morales, A. et al., Learning to predict grasp reliability for a multifinger robot hand by using visual features, *Proceedings of the Eighth IASTED International Conference on Artificial Intelligence and Soft Computing*, 2004, pp. 249–254.
241. Morales, A. et al., An active learning approach for assessing robot grasp reliability, *Proceedings of the IEEE/RSJ International Conference on Intelligent Robots and Systems (IROS)*, Vol. 1, 2004, pp. 485–490.
242. Morita, T., Sugano, S., A technology map for standardizing safety measurements of human symbiotic robots, *Advanced Robotics*, Vol. 13, No. 3, 1999, pp. 307–308.
243. Morita, T., Sugano, S., Double safety measure for human symbiotic manipulator, *Proceedings of the IEEE/ASME International Conference on Advanced Intelligent Mechatronics*, 1997, pp. 130–131.
244. Motamed, C., Schmitt, A., Vision based safety device for uncontrolled robotic environments, *Proceedings of the IEEE International Conference on Systems, Man, and Cybernetics*, 1993, pp. 528–533.
245. Muldau, H.H.V., Safety at work-place using industrial robots, *Proceedings of the 8th International Symposium on Industrial Robots*, 1978, pp. 210–215.
246. Murakami, M., A safety failover subsystem for intelligent mobile robots, *Proceedings of the International Conference on Control, Automation and Systems*, 2007, pp. 2493–2498.
247. Murakami, M., Development of a duplex computer system for humanoid robot applications: Design of the safety failover subsystem, *Proceedings of the 33rd Annual Conference of the IEEE Industrial Electronics Society*, 2007, pp. 2783–2788.
248. Murakami, M., Fault tolerance design for computers used in humanoid robots, *Proceedings of the International Conference on Intelligent Robots and Systems*, 2007, pp. 2301–2307.
249. Musto, J.C., Saridis, G.N., Entropy-based reliability analysis for intelligent machines, *IEEE Transactions on Systems, Man, and Cybernetics, Part B: Cybernetics*, Vol. 27, No. 2, 1997, pp. 239–244.
250. Myers, J.K., Agin, G.J., A supervisory collision avoidance system for robot controllers, *Proceedings of the Annual Meeting of American Society of Mechanical Engineers*, 1982, pp. 59–63.
251. Myklebust, E., Thiel, E., Practical robot safety systems, *Proceedings of the 16th International Symposium on Industrial Robots*, 1986, pp. 1137–1146.

252. Nagamachi, M., Anayama, Y., An ergonomic study of the industrial robot, Part I: The experiments of unsafe behavior on robot manipulation, *Japanese Journal of Ergonomics*, Vol. 19, 1983, pp. 259–264.

253. Nagamachi, M. et al., A human factors study of the industrial robot, Part II: Human reliability on robot manipulation, *Japanese Journal of Ergonomics*, Vol. 20, 1984, pp. 55–64.

254. Nagamachi, M., Two fatal accidents due to robots in Japan, in *Ergonomics of Hybrid Automated System I*, ed. W. Karwowski, Elsevier, Amsterdam, 1988, pp. 391–396.

255. Narita, M., Shimamura, M., Oya, M., Reliable protocol for robot communication on web services, *Proceedings of the International Conference on Cyberworlds*, 2005, pp. 210–220.

256. Narita, S., Ohkami, Y., Development of distributed controller software for improving robot performance and reliability, *Proceedings of the IEEE/RSJ International Conference on Intelligent Robots and Systems (IROS)*, Vol. 3, 2004, pp. 2384–2389.

257. Navarro-Alarcon, D., Parra-Vega, V., Olguin-Diaz, E., Minimum set of feedback sensors for high performance decentralized cooperative force of redundant manipulators, *Proceedings of the IEEE International Workshop on Robotic and Sensors Environments*, 2008, pp. 114–119.

258. Newkirk, J.T., Bowling, A.P., Renaud, J.E., Workspace characterization of a robotic system using reliability-based design optimization, *Proceedings of the IEEE International Conference on Robotics and Automation*, 2008, pp. 3958–3963.

259. Ng., W.S., Tan, C.K., On safety enhancements for medical robots, *Reliability Engineering and System Safety*, Vol. 54, 1996, pp. 35–45.

260. Nicolaisen, P., The development of safety devices to fit the problems-example: Industrial robots, *Industrie Robotor*, Vol. 102, No. 73, 1980, pp. 164–169.

261. Noro, K., Okada, Y., Robotization and human factors, *Ergonomics*, Vol. 26, 1983, pp. 985–1000.

262. Ntuen, C.A., Park, E.H., Formal method to characterize robot reliability, *Proceedings of the Annual Reliability and Maintainability Symposium*, 1993, pp. 395–397.

263. Okina, S. et al., Study of a self-diagnosis system for an autonomous mobile robot, *Advanced Robotics*, Vol. 14, No. 5, 2000, pp. 339–341.

264. Olex, M.B., Shulman, H.G., Human factors efforts in robotics system design, *Proceedings of the 13th International Symposium on Industrial Robots*, 1983, pp. 9.29–9.36.

265. Otsuka, K., Inspection robot system of electronic unit manufacturing, *Robot*, No. 111, 1996, pp. 22–27.

266. Parker, L.E., Draper, J.V., Robotics applications in maintenance and repair, in *Handbook of Industrial Robotics*, ed. S. Y. Nof, John Wiley and Sons, New York, 1999, pp. 1023–1036.

267. Parsons, M.H., Robot system safety issues best considered in design phase, *Occupational Health and Safety*, Vol. 53, No. 8, 1984, pp. 38–42.

268. Patooghy, A. et al., A solution to single point of failure using voter replication and disagreement detection, *Proceedings of the 2nd IEEE International Symposium on Dependable, Automatic and Secure Computing*, 2006, pp. 171–176.

269. Pegman, G., Reed, J., Safety and usability of telerobotic systems, *Proceedings of the International Conference on Nuclear Engineering*, 1997, pp. 1935–1941.

270. Percival, N., Safety aspects of robots and flexible manufacturing systems, *Proceedings of the 1st International Conference on Human Factors in Manufacturing*, 1984, pp. 179–183.

271. Piggin, R., A new approach to robotic safety: Safety BUS p at BMW, *Industrial Robot*, Vol. 29, No. 6, 2002, pp. 524–529.

272. Pirjanian, P., Reliable reaction, *Proceedings of the IEEE/SICE/RSJ International Conference on Multisensor Fusion and Integration for Intelligent Systems*, 1996, pp. 158–165.

273. Polakoff, P.L., Man's marriage to robotics: A "For Better or Worse" Union, *Occupational Health and Safety*, Vol. 54, No. 4, 1985, pp. 24–25.

274. Pollard, B.W., RAM for robots: Reliability, availability, maintainability, *Robotics Today*, 1980, pp. 51–53.

275. Pomerleau, D.A., Reliability estimation for neural network based autonomous driving, *Robotics and Autonomous Systems*, Vol. 12, No. 3–4, 1994, pp. 113–119.

276. Potter, R.D., Safety for robotics, *Proceedings of the American Society of Safety Engineers Professional Development Conference*, 1983, pp. 171–183.

277. Pouliezos, A., Stavrakakis, G.S., Fast fault diagnosis for industrial processes applied to the reliable operation of oobotic systems, *International Journal of Systems Science*, Vol. 20, 1989, pp. 1233–1257.

278. Pruski, A., Surface contact sensor for robot safety, *Sensor Review*, Vol. 6, No. 3, 1986, pp. 143–144.

279. Rachkov, M., Safety systems of technological climbing robots, *Proceedings of the IEEE International Symposium on Industrial Electronics*, 1997, pp. 660–665.

280. Rahimi, M., Karwowski, W., Research paradigm in human-robot interaction, *International Journal of Industrial Ergonomics*, Vol. 5, No. 1, 1990, pp. 59–71.

281. Rahimi, M., System safety approach to robot safety, *Proceedings of the 28th Human Factors Society Annual Meeting*, 1984, pp. 102–106.

282. Rahimi, M., System safety for robots: Ali energy barrier analysis, *Journal of Occupational Accidents*, Vol. 8, 1986, pp. 127–138.

283. Rahimi, M., Xiadong, X., Framework for software safety verification of industrial robot operations, *Computers and Industrial Engineering*, Vol. 20, No. 2, 1991, pp. 279–287.

284. Ramachandran, S., Nagarajan, T., Sivaprasad, N., Reliability studies on assembly robots using the finite element method, *Advanced Robotics*, Vol. 7, No. 4, 1993, pp. 385–393.

285. Ramachandran, V., Vajpayee, S., Safety in robotic installations, *Robotics, and Computer Integrated Manufacturing*, Vol. 3, No. 3, 1987, pp. 301–308.

286. Ramirez, C.A., Artificial intelligence applied to robot fail safe operations, *Proceedings of the Robot 9 Conference on Current Issues and Future Concerns*, 1985, pp. 9.21–9.37.

287. Ramirez, C.A., Safety of robot, in *Handbook of Industrial Robots*, ed. S.Y. Nof, John Wiley and Sons, New York, 1985, pp. 131–148.

288. Rao, S.S., Bhatti, P.K., Probabilistic approach to manipulator kinematics and dynamics, *Reliability Engineering and System Safety*, Vol. 72, 2001, pp. 47–58.

289. Reinkensmeyer, D.J., Wolbrecht, E., Bobrow, J., A computational model of human-robot load sharing during robot-assisted arm movement training after stroke, *Proceedings of the 29th Annual International Conference of the IEEE Engineering in Medicine and Biology Society*, 2007, pp. 4019–4023.

290. Renaud, P., Cervera, E., Martinet, P., Towards a reliable vision-based mobile robot formation control, *Proceedings of the IEEE/RSJ International Conference on Intelligent Robots and Systems*, Vol. 4, 2004, pp. 3176–3181.

291. Roston, G.P., Dowling, K., Drivetrain design, incorporating redundancy, for an autonomous walking robot, *Proceedings of the ASCE Specialty Conference on Robotics for Challenging Environments Conference*, 1994, pp. 184–192.

292. Rovetta, A., Telerobotic surgery control and safety, *Proceedings of the IEEE International Conference on Robotics and Automation*, 2000, pp. 2895–2900.

293. Rudall, R.H., Automation and robotics world-wide reports and surveys, *Robotica*, Vol. 14, 1996, pp. 243–151.

294. Russel, J., Robot safety considerations: A check list, *Professional Safety*, Vol. 28, 1983, pp. 36–37.

295. Russell, R.A. et al., A robotic system to locate hazardous chemical leaks, *Proceedings of the IEEE International Conference on Robotics and Automation*, 1995, pp. 556–561.

296. Rybski, P. et al., Sensor fusion for human safety in industrial work-cells, *Proceedings of the IEEE/RSJ International Conference on Intelligent Robots and Systems*, 2012, pp. 1–8.

297. Sakai, H., Amasaka, K., The robot reliability design and improvement method and the advanced Toyota production system, *Industrial Robot*, Vol. 34, No. 4, 2007, pp. 310–316.

298. Salini, D., Reliability prediction with Mil handbook and field data collection in FMS, and robotics, *Microelectronics and Reliability*, Vol. 29, No. 3, 1989, pp. 415–418.

299. Salvendy, G., Review and appraisal of human aspects in planning robotic systems, *Behavior and Information Technology*, Vol. 2, 1983, pp. 262–287.

300. Sanderson, A.C., Modular robotics: Design and example, *Proceedings of the IEEE Conference on Emerging Technologies and Factory Automation*, Vol. 2, 1996, pp. 460–466.

301. Sanderson, L.M., Collins, J.W., McGlothlin, J.D., Robot-related fatality involving a US manufacturing plant employee: Case report and recommendations, *Journal of Occupational Accidents*, Vol. 8, 1986, pp. 13–23.

302. Sato, Y., Inone, K., Kumamoto, H., The safety assessment of the human–robot systems, *Bulletin of JSME*, Vol. 29, No. 250, 1980, pp. 1356–1361.

303. Sato, Y., Inone, K., Kumamoto, H., The safety assessment of the human–robot systems, *Bulletin of JSME*, Vol. 29, No. 256, 1986, pp. 3618–3525.

304. Sato, Y., Inone, K., Kumamoto, H., The safety assessment of the human–robot systems, *Bulletin of JSME*, Vol. 29, No. 257, 1986, pp. 3645–3651.

305. Sato, Y., Inoue, K., Heneley, E.J., Action chain model for the design of the hazard control systems for robots, *IEEE Transactions on Reliability*, Vol. 39, No. 2, 1990, pp. 151–157.

306. Sato, Y., Inoue, K., Heneley, E.J., Hazard assessment of industrial robots, *Proceedings of the USA-Japan Symposium on Flexible Automation*, 1988, pp. 703–708.

307. Savsar, M., Reliability analysis of a flexible manufacturing cell, *Reliability Engineering & System Safety*, Vol. 67, No. 2, 2000, pp. 147–152.

308. Schneider, D.L., Tesar, D., Barnes, J.W., Development and testing of a reliability performance index for modular robotic systems, *Proceedings of the Annual Reliability and Maintainability Symposium*, 1994, pp. 263–271.

309. Schreiber, R.R., Robot safety: A shared responsibility, *Robotics Today*, 1983, pp. 61–65.
310. Seim, B., Beutler, B., Stringent safety guidelines bring safeguarding devices into focus, *Robotics World*, Vol. 11, No. 3, 1993, pp. 16–18.
311. Seward, D., Margrave, F., LUCIE the robot excavator-design for system safety, *Proceedings of the IEEE International Conference on Robotics and Automation*, 1996, pp. 963–968.
312. Sharma, S.P., Kumar, D., Kumar, A., Reliability analysis of complex multi-robotic system using GA and fuzzy methodology, *Applied Soft Computing*, Vol. 12, 2012, pp. 405–415.
313. Sharp, R., Decreton, M., Radiation tolerance of components and material in nuclear robot applications, *Reliability Engineering and System Safety*, Vol. 53, 1996, pp. 291–299.
314. Sheehy, N.P., Chapman, A.J., Safety of CNC and robot technology, *Journal of Occupational Accidents*, Vol. 10, No. 1, 1988, pp. 21–28.
315. Shi, T. et al., Approach to fault on-line detection and diagnosis based on neural networks for robot in FMS, *Chinese Journal of Mechanical Engineering (English Edition)*, Vol. 11, No. 2, 1998, pp. 115–121.
316. Shima, D., Safety control on introduction of industrial robots to factories, *Safety*, Vol. 33, 1982, pp. 18–19.
317. Shulman, H.G., Olex, M.B., Designing the user-friendly robot, *Human Factors*, Vol. 27, 1985, pp. 91–98.
318. Smrcek, J., Neupauer, R., Testing of intelligent robots, development and experience, *Proceedings of the IEEE International Conference on Intelligent Engineering Systems*, 1997, pp. 119–121.
319. Sneckenberger, J.E., Etherton, J.R., Computer related hazard control needs for safe robot operations, *Proceedings of the American Society of Mechanical Engineers (ASME) International Computers in Engineering Conference*, 1983, pp. 110–113.
320. Stancliff, S.B., Dolan, J., Trebi-Ollennu, A., Planning to fail-reliability needs to be considered a priori in multirobot task allocation, *Proceedings of the IEEE International Conference on Systems, Man, and Cybernetics*, 2009, pp. 2362–2367.
321. Stancliff, S.B., Dolan, J.M., Planning to fail-using reliability to improve multirobot task allocation, *Proceedings of the SPIE Conference on Unattended Ground, Sea, and Air Sensor Technologies and Applications*, Vol. 7693, 2010, pp. 1–9.
322. Stormont, D.P., Allan, V.H., Managing risk in disaster scenarios with autonomous robots, *Journal of Systemics, Cybernetics & Informatics*, Vol. 7, No. 4, 2009, pp. 66–71.
323. Stowe, W.W., Robots, safe or hazardous?, *Professional Safety*, Vol. 28, 1983, pp. 32–35.
324. Stubbs, K., Hinds, P., Wettergreen, D., Challenges to grounding in human-robot collaboration: Errors and miscommunications in remote exploration robotics, Report No. CMU-RI-TR-06-32, Robotics Institute, Carnegie Mellon University, Pittsburgh, Pennsylvania, July 2006, pp. 1–17.
325. Sugimoto, N., Introduction to safety measures for industrial robots, *Safety*, Vol. 33, 1982, pp. 14–17.
326. Sugimoto, N., Kawaguchi, K., Fault tree analysis of hazards created by robot, *Proceedings of the 13th International Symposium on Industrial Robots and Robot 7*, 1983, pp. 9.13–9.28.

327. Sugimoto, N., Systematic robot related accidents and standardization of safety measures for robots, *Proceedings of the 14th International Symposium on Industrial Robots*, 1984, pp. 131–138.

328. Suita, K. et al., Failure-to-safety "Kyozon" system with simple contact detection and stop capabilities for safe human autonomous robot coexistence, *Proceedings of the IEEE International Conference on Robotics and Automation*, 1995, pp. 3089–3096.

329. Sun, U., Sneckenberger, J., Human-robot symbotic system safety "Criteria" for manned space flight, *Proceedings of the Annual Pittsburgh Conference: Control, Robotics, Systems, Power and Mechanical Modeling and Simulation*, Vol. 21, 1992, pp. 2327–2335.

330. Suri, R., Quantitative techniques for robotic systems analysis, in *Handbook of Industrial Robotics*, ed. S.Y. Nof, John Wiley and Sons, New York, 1985, pp. 605–638.

331. Tadokoro, S. et al., Stochastic prediction of human motion and control of robots in the service of human, *Proceedings of the International Conference on Systems, Man and Cybernetics*, 1993, pp. 503–508.

332. Takeda, T., Hirata, Y., Kosuge, K., HMM-based error recovery of dance step selection for dance partner robot, *Proceedings of the IEEE International Conference on Robotics and Automation*, 2007, pp. 1768–1773.

333. Taogeng, Z., Dingguo, S., Qun, H., Application of Bayes data fusion in robot reliability assessment, *Proceedings of the International Workshop on Bio-Robotics and Teleoperation*, 2001, pp. 262–265.

334. Tian, J., Sneckenberger, J., Performance evaluation of three pressure mats as robot work-station safety devices, in *Ergonomics of Hybrid Automated Systems*, ed. W. Karwowski, Elsevier, Amsterdam, 1988, pp. 559–565.

335. Ting, Y., Tosunoglu, S., Fernandez, B., Control algorithm for fault tolerant robots, *Proceedings of the IEEE International Conference on Robotics and Automation*, 1994, pp. 910–915.

336. Tinos, R., Navarro-Serment, L.E., Paredis, C.J.J., Fault tolerant localization for teams of distributed robots, *Proceedings of the International Conference on Intelligent Robot and Systems*, 2001, pp. 1061–1066.

337. Toye, G., Leifer, L.J., Helenic fault tolerance for robots, *Computers & Electrical Engineering*, Vol. 20, No. 6, 1994, pp. 479–497.

338. Troccaz, J., Passive arm with dynamic constraints: A solution to safety problems in medical robotics?, *Proceedings of the IEEE International Conference on Systems, Man, and Cybernetics*, 1993, pp. 166–171.

339. Tso, K., Backes, P., Fail-safe tele-autonomous robotic system for nuclear facilities, *Robotics and Computer-Integrated Manufacturing*, Vol. 10, No. 6, 1993, pp. 423–428.

340. Ukidve, C.S., McInroy, J.E., Jafari, F., Using redundancy to optimize manipulability of Stewart platforms, *IEEE/ASME Transactions on Mechatronics*, Vol. 13, No. 4, 2008, pp. 475–479.

341. Van Deest, R., Robotic safety: A potential crisis, *Professional Safety*, Vol. 29, 1984, pp. 40–42.

342. Vanderperre, E.J., Makhanov, S.S., Risk analysis of a robot-safety device system, *International Journal of Reliability, Quality and Safety Engineering*, Vol. 9, No. 1, 2002, pp. 79–87.

343. Wakita, Y. et al., Realization of safety in a coexistent robotic system by information sharing, *Proceedings of the IEEE International Conference on Robotics and Automation*, 1998, pp. 3474–3479.
344. Walker, I.D., Cavallaro, J.R., The use of fault-trees for the design of robots for hazardous environments, *Proceedings of the Annual Reliability and Maintainability Symposium*, 1996, pp. 229–235.
345. Wang, T. et al., Fuzzy reliability design of robot parts based on Weibull and normal distributions, *Proceedings of the International Conference on Computer Science and Software Engineering*, 2008, pp. 1081–1084.
346. Ward, G.R., Safety in robot control systems, *Measurement and Control*, Vol. 21, No. 9, 1988, pp. 266–271.
347. Wei, Y. et al., Visual error augmentation for enhancing motor learning and rehabilitative relearning, *Proceedings of the International Conference on Rehabilitation Robotics*, 2005, pp. 505–510.
348. Wells, D.J., Hazardous area robotics for nuclear systems maintenance: A challenge in reliability, *Proceedings of the Portland International Conference on Management of Engineering and Technology*, 1991, pp. 371–373.
349. Wells, D.J., Krishnaswami, K., Fault analysis and recovery strategies for deep-sea robots, *Proceedings of the 11th Annual Energy Sources Technology Conference and Exhibition*, 1988, pp. 39–47.
350. Wewerinke, P.H., Modeling human operator involvement in robotic systems, *Proceedings of the IEEE International Conference on Systems*, 1991, pp. 1225–1230.
351. Wikman, T.S., Branicky, M.S., Newman, W.S., Reflex control for robot system preservation, reliability and autonomy, *Computers & Electrical Engineering*, Vol. 20, No. 5, 1994, pp. 391–407.
352. Wikman, T.S., Newman, W.S., Reflex control for safe autonomous robot operation, *Reliability Engineering and System Safety*, Vol. 53, 1996, pp. 339- 347.
353. Willetts, N., *Jaguar. A* robot user's experience, *Industrial Robot*, Vol. 20, No. 1, 1993, pp. 9–12.
354. Wilson, M.S., Reliability and flexibility: A mutually exclusive problem for robotic assembly?, *IEEE Transactions on Robotics and Automation*, Vol. 12, No. 2, 1996, pp. 343–347.
355. Winfield, A.F.T., Nembrini, J., Safety in numbers: Fault-tolerance in robot swarms, *International Journal of Modelling, Identification and Control*, Vol. 1, No. 1, 2006, pp. 30–37.
356. Wygant, R.M., Donaghey, C.E., Ergonomic considerations in robot selection and safety, *Proceedings of the 31st Annual Meeting of the Human Factors Society*, 1987, pp. 181–185.
357. Xu, F., Deng, Z., The reliability apportionment for X-ray inspection real time imaging pipeline robot based on fuzzy synthesis, *Proceedings of the Fifth World Congress on Intelligent Control and Automation*, 2004, pp. 4763–4767.
358. Yamada, A., Takata, S., Reliability improvement of industrial robots by optimizing operation plans based on deterioration evaluation, *CIRP Annals–Manufacturing Technology*, Vol. 51, No. 1, 2002, pp. 319–322.
359. Yamada, Y. et al., Failure-to-safety robot system for human–robot coexistence, *Robotics and Autonomous Systems*, Vol. 18, No. 1–2, 1996, pp. 283–291.
360. Yamada, Y. et al., FTA-based issues on securing human safety in human/robot coexistence system, *Proceedings of the IEEE International Conference on Systems, Man, and Cybernetics*, 1999, pp. 1058–1063.

361. Yamada, Y., Evaluation of human pain tolerance and its application to designing safety robot mechanism for human–robot coexistence, *Journal of Robotics and Mechatronics*, Vol. 9, 1997, pp. 65–70.

362. Yang, M. et al., Environmental modeling and obstacle avoidance of mobile robots based on laser radar, *Qinghua Daxue Xuebao/Journal of Tsinghua University*, Vol. 40, No. 7, 2000, pp. 112–116.

363. Yeh, P., Barash, S., Wysocki, E., Vision system for safe robot operation, *Proceedings of the IEEE International Conference on Robotics and Automations*, 1988, pp. 1461–1465.

364. Yokomizo, Y., Occupational and safety problems caused by increases in the robot population, in *Occupational Health and Safety in Automation and Robotics*, ed. K. Noro, Taylor & Francis, London, 1987, pp. 169–174.

365. Yong, J. et al., A combined logistic and model-based approach for fault detection and identification in a climbing robot, *Proceedings of the IEEE International Conference on Robotics and Biomimetics*, 2006, pp. 1512–1516.

366. Yu, S., McCluskey, E.J., On-line testing and recovery in TMR systems for real-time applications, *Proceedings of the International Test Conference*, 2001, pp. 240–249.

367. Zhang, C., Bai, G., Extremum response surface method of reliability analysis on two-link flexible robot manipulator, *Journal of Central South University of Technology*, Vol. 19, 2012, pp. 101–107.

368. Zhang, J., Lu, T., Fuzzy fault tree analysis of a cable painting robot, *Shanghai Jiaotong Daxue Xuebao/Journal of Shanghai Jiaotong University*, Vol. 37, No. 5, 2003, pp. 62–65.

369. Zhang, X. et al., Experimental analysis of in-pipe robot accelerated life, *Proceedings of the International Conference on Mechatronic Science*, 2011, pp. 572–575.

370. Zhongxiu, S., Reliability analysis and synthesis of robot manipulators, *Proceedings of the Annual Reliability and Maintainability Symposium*, 1994, pp. 201–205.

371. Zhou, G., Position and orientation calibration and moving reliability of the robot used in pyrometallurgy process, *Zhongnan Gongye Daxue Xuebao/Journal of Central South University of Technology*, Vol. 31, No. 6, 2000, pp. 556–560.

372. Zurada, J.G., James, H., Sensory integration in a neural network-based robot safety system, *International Journal of Human Factors in Manufacturing*, Vol. 5, No. 3, 1995, pp. 325–340.

Index

A

Accident, 5, 115; *see also* Robot-related accidents
  factors causing, 53
  in robots operation, 120
  TOR method use, 68
Accident-causation theories, 51
  Domino, 52–53
  human factors, 51–52
AC voltmeter, *see* Alternating current voltmeter (AC voltmeter)
Adjusting tools for changeovers cost (ATCC), 178
Alternating current voltmeter (AC voltmeter), 135
American National Standards Institute (ANSI), 8, 77
ANSI, *see* American National Standards Institute (ANSI)
Arithmetic mean, 13–14
ATCC, *see* Adjusting tools for changeovers cost (ATCC)
Awareness barrier, 4
Axioms of Industrial Safety, 52–53

B

Barrier, 4; *see also* Energy barrier analysis
  awareness, 4
  physical, 105–106
Basic events, *see* Primary events
Bathtub hazard rate curve, 31, 32
  burn-in period, 31, 32
  distribution, 22–23
  equation, 32–33
  useful-life period, 32
  wear-out period, 32
Behavioral and organizational factors, 118
Boolean algebra, 13
  laws, 15–16
Breaking injuries, 49
Bridge network, 45–46

C

Casual observers, 159
Conference proceedings, 6
Continuous task, 5
Control errors, 99
Control subsystem's algorithm, 100
Crushing injuries, 49
Cumulative distribution function, 18–19, 143
Cutting injuries, 49

D

Data sources, 7–8
  for reliability, 185–187
DC, *see* Direct current (DC)
Decreasing-hazard-rate region, 32
Deviation analysis, 162
Dhillon distribution, 22
Direct current (DC), 90, 135
Direct labor cost (DLC), 178
Distribution probability density function, 22
DLC, *see* Direct labor cost (DLC)
"Domino" accident-causation theory, 52–53
Downtime, 4
  robot, 79, 137

E

Electric robot, reliability analysis of, 90; *see also* Hydraulic robot, reliability analysis of
  assumptions/factors, 90
  block diagram, 91
  example, 93
  independent parts, 92
  subsystems, 91
Electrical hazards, 48, 111
Electrical safety features, 102
Electrical/electronic subsystem, 100
End-effector, 4

Energy barrier analysis, 162–163
Energy hazards, 48–49
Engineering factors, 118
Engineering/maintenance department, 100–101
Environmental hazards, 48
Erratic robot, 4
Error recovery, 4
Event data banks, 186
Expected value, 19–20
Exponential distribution, 20–21

F

FACE, *see* Nebraska Fatality Assessment and Control Evaluation (FACE)
Factory integration process, 153
Fail-safe, 4
Failure Analysis, 58
Failure density function, *see* Probability density function
Failure Mode, Effects, and Criticality Analysis (FMECA), 58
Failure modes and effect analysis (FMEA), 57–59, 69, 85
Failure Reporting, Analysis, and Corrective Action (FRACAS), 8
Fault tree analysis (FTA), 57, 63, 85, 125, 163–165; *see also* Interface safety analysis
　advantages and disadvantages, 67
　application, 125
　construction, 63
　dark robot system room, 65
　event occurrence probability values, 67
　examples, 64, 65, 125–127
　OR and AND gates, 64
　probability evaluation, 65–67
　top event, 126
Field failure data collection, 187
Flashing lights, 106
Flow relationships, 70
FMEA, *see* Failure modes and effect analysis (FMEA)
FMECA, *see* Failure Mode, Effects, and Criticality Analysis (FMECA)
Ford method, 173–174

FRACAS, *see* Failure Reporting, Analysis, and Corrective Action (FRACAS)
FTA, *see* Fault tree analysis (FTA)

G

Government/Industry Data Exchange Program (GIDEP), 7
Gripper subsystem, 4, 87

H

Hazard and Operability analysis (HAZOP analysis), 69, 70
Hazard rate
　function, 34
　robot, 84
Hazardous motion, 4
HAZOP analysis, *see* Hazard and Operability analysis (HAZOP analysis)
Highways, 147–148
Hindu-Arabic numeral system, 13
Human accidents, robot-related, 159
Human errors, 78, 99
Human factors, 153
　accident-causation theory, 51–52
　communication, 156
　hazards, 48
　humans at risk from robots, 159
　issues during robotic system, 153
　job design, 156
　limiting robot movements, 161
　maintainability, 154–155
　management issues, 156
　operators, guidelines for safeguarding robot, 160–161
　risk-reducing measurement, 159
　robotization advantages and disadvantages, 157–159
　safety, 154
　selection and training, 155
　teachers, guidelines for safeguarding robot, 160
　work environment, 155
　worker–machine interface, 155

Hydraulic robot, reliability analysis
    of, 86; *see also* Electric robot,
    reliability analysis of
  assumptions/factors, 86
  block diagram, 86
  example, 89–90
  independent parts, 88
  subsystems, 87

I

Inappropriate activities factor, 52
Inappropriate response/incompatibility
    factor, 51
Institute for Production and
    Automation (IPA), 174
Intelligent systems, 106
Interface safety analysis, 69
  flow relationships, 70
  functional relationships, 69
  physical relationships, 71
International Organization for
    Standardization (ISO), 8
IPA, *see* Institute for Production and
    Automation (IPA)
IPA–Stuttgard method, 174
ISO, *see* International Organization for
    Standardization (ISO)

K

k-out-of-n network, 41–43
Kinematic hazards, 48

L

Laboratory tests, 185
Laplace transforms, 13, 23, 24, 193
  final value theorem, 25
  first-order differential equations, 25–27
  of functions, 24
Laser welding, 109–110
Life cycle costing, 182

M

Maintainability, robot, 154–155
Maintenance personnel, 159

Management, 98, 101, 156
Markov method, 59, 85, 127, 165–167, 192,
    196, 199, 202
  assumptions, 59–60
  examples, 60–63, 127–130
  robot system state-space diagram,
    60, 128
Mathematical modeling, 191
  model I, 191–194
  model II, 195–198
  model III, 198–201
  model IV, 201–204
  model V, 205–207
  model VI, 207–209
  robot state-space diagram, 192, 198,
    202, 205
  robot-related reliability and safety, 191
Mathematics, 13
  arithmetic mean, 13–15
  Boolean algebra laws, 15–16
  Laplace transform, 23–25
  mean deviation, 13–15
  probability definition and properties,
    16–17
  probability distribution-related
    definitions, 18–20
  probability distributions, 20–23
  problems, 27
MC, *see* Miscellaneous cost (MC)
Mean deviation, 14–15
Mean time to failure, 35–36
  bridge network, 46
  k-out-of-n network/system, 43
  series-system, 38
  standby system, 44
Mean time to robot-related problems
    (MTTRP), 79, 80
Mean time to robot failure (MTTRF), 4,
    62, 79, 80, 81–82, 129, 200
Mechanical hazards, 99, 111
Mechanical injuries, 48–49
Mechanical safety features, 101
Mechanical subsystem, 99–100
Mechanical-related problems, 99
Minimum cost rule, 183–184
Miscellaneous cost (MC), 178
Mission time, 4, 83
Misuse-and-abuse hazards, 48

MTTRF, *see* Mean time to robot failure (MTTRF)
MTTRP, *see* Mean time to robot-related problems (MTTRP)

N

n-unit system, 37
National Bureau of Standards (NBS), 174
National Institute of Occupational Safety and Health (NIOSH), 7, 163
National Technical Information Service (NTIS), 7, 8
NBS, *see* National Bureau of Standards (NBS)
Nebraska Fatality Assessment and Control Evaluation (FACE), 7
NIOSH, *see* National Institute of Occupational Safety and Health (NIOSH)
NTIS, *see* National Technical Information Service (NTIS)
Nuclear industry, 146–147

O

Operational and control subsystems' software, 100
Operational procedures, 100
Organizations, 9

P

Parallel network, 39–41
Parts count method, 85
Payback period, 182–183
Pendant, 4
Periodic inspection records, 187
Physical barriers, 105–106
Physical relationships, 71
Power line maintenance, 149
Primary events, 63
Probability
    cumulative distribution function, 18–19
    definition and properties, 16–17
    density function, 19, 20, 21, 33
    distribution-related definitions, 18–20

expected value, 19–20
    tree method, 71–73
Probability distributions, 20
    Bathtub hazard rate curve distribution, 22–23
    exponential distribution, 20–21
    Rayleigh distribution, 21–22
    Weibull distribution, 22
Product hazards, 48–49
Product safety, organization tasks for, 49–50
Puncturing injuries, 49

R

RADC, *see* Rome Air Development Center (RADC)
Railways, 148
Random failure, 4
    component failures, 77–78
Rayleigh distribution, 21–22
RCA, *see* Root cause analysis (RCA)
Redundancy, 4, 79
Reliability, 4
    accident-causation theories, 51–53
    bathtub hazard rate curve, 31–33
    bridge network, 45–46
    configurations, 37
    data sources for, 185–187
    function, 34–35
    k-out-of-n network, 41–43
    mechanical injuries, 48–49
    need for safety and role of engineers, 47–48
    parallel network, 39–41
    problems, 54
    product hazards, 48–49
    and safety basics, 31
    series network, 37–39
    standby system, 43–45
Reliability-related formulas, 33
    hazard rate function, 34
    mean time to failure, 35–36
    probability density function, 33
    reliability function, 34–35
Return on investment, 184
Risk priority number (RPN), 58
Robot accidents and analysis, 115
    causes and sources, 117–118

effects, 118–119
examples, 115–117
fault tree analysis, 125–127
General Motors Corporation
    study, 118
Japanese study, 117
at manufacturer and user facilities,
    119–120
Markov method, 127–130
methods for, 122–130
probability estimation of accident
    occurrence, 122–123
problems, 130
RCA, 123–125
recommendations to human injury
    prevention, 120–122
robot system design, 121
supervision of workers, 121
training of workers, 121–122
Robot costing, 171
annual corrective maintenance cost
    estimation, 179
investment cost estimation, 178–179
life cycle cost estimation models,
    181–182
models for estimating cost, 178
operating cost estimation, 178
possible cost of losses estimation, 180
robotic installation and start-up cost
    estimation, 180
unit cost of assembly estimation, 179
Robot engineering cost (RC$_e$), 178
Robot life cycle cost estimation models,
    181–182
Robot maintenance-related needs, 133
areas of robotics applications, 145–149
maintenance types, 133
measuring instruments, 134–135
models for, 137–145
safeguard robot maintenance
    personnel, 136–137
tooling for periodic robot
    inspections, 134–135
Robot operating cost (ROC), 178
Robot performance testing, 171–173
Ford method, 173–174
IPA–Stuttgard method, 174
NBS, 174
test-robot program method, 173

Robot procurement cost (RC$_p$), 178
Robot reliability, *see* Robot system
    reliability (RSR)
Robot safety, 97
causes of hazards, 99
features and functions, 106–107
hazards, 97–99
problems, 97–99, 111–112
robot safeguard approaches, 105–106
robot-related hazards, 98
robotized welding operations, safety
    considerations for, 107–111
roles of robot manufacturers and
    users, 99–101
weak points in planning, design, and
    operation, 104–105
Robot special tooling cost (RC$_{st}$), 178
Robot system mean time to failure
    (RSMTTF), 200, 204, 209
Robot system reliability (RSR), 77, 82–83,
    200, 203
analysis methods and models, 84–85
analysis of hydraulic and electric
    robots, 86–93
categories, 175
causes and corrective measures, 77–79
guidelines for, 176
management-related tasks in, 175
problems, 93–94
robot effectiveness dictating
    factors, 79
robot failures, 77–79
robot-related reliability measures,
    80–84
survey results, 79
testing, 174
Robot-related accidents
effects of, 118–119
examples, 115–117
at manufacturer and user facilities,
    119–120
Robot-related hazards, 98
Robot-related human tasks, 156–157
robot tasks, 157
Robot-related reliability measures, 80
MTTRF, 81–82
MTTRP, 80
robot reliability, 82–83
robot's hazard rate, 84

Robot(s) (robot systems), 1, 3–5
    books, 6–7
    conference proceedings, 6
    data sources, 7–8
    design phase, 101–102
    diagnosis and monitoring approach,
        135
    effectiveness dictating factors, 79
    failures, 77–79, 171, 184–185
    FMEA, 57–59
    FTA, 63–67
    hazard rate, 84
    HAZOP analysis, 69
    installation phase, 102
    interface safety analysis, 69–71
    journals, 5–6
    Markov method, 59–63
    mean time to failure, 4
    mean time to repair, 4
    operation and maintenance
        phase, 103
    organizations, 9
    powering system, 133
    probability tree method, 71–73
    problems, 10, 73–74
    programmers, 159
    programming phase, 103
    project cost, 177–178
    repair and inspection records-related
        requirements for, 187–188
    standards, 8–9
    system design, 121
    system reliability/safety-related
        facts, 2–3
    technical reports, 7
    testing, 171, 176–177
    TOR, 68
    useful sources, 5–9
Robotics
    deviation analysis, 162
    energy barrier analysis, 162–163
    fault tree analysis, 163–165
    Markov method, 165–167
    safety and human error analysis
        in, 162
    task analysis, 163
Robotization
    life cycle costing, 182
    minimum cost rule, 183–184

    payback period, 182–183
    return on investment, 184
Robotization project investment cost
        (RPIC), 178
Robotized gas-shielded arc
        welding, 110
Robotized resistance welding, 111
Robotized welding operations, safety
        considerations for, 107
    factors, 108
    robotized gas-shielded arc welding,
        110
    robotized laser welding hazards,
        109–110
    robotized resistance welding, 111
ROC, *see* Robot operating cost (ROC)
Rome Air Development Center
        (RADC), 7
Root cause analysis (RCA), 123
    advantages and disadvantages,
        124–125
    performance, 123
    software, 124
    steps, 124
RPIC, *see* Robotization project
        investment cost (RPIC)
RPN, *see* Risk priority number (RPN)
RSMTTF, *see* Robot system mean time to
        failure (RSMTTF)
RSR, *see* Robot system reliability (RSR)

S

Safeguard, 4
Safety, 4, 154; *see also* Robot safety
    department, 101
Safety management principles, 50–51
Safety Research Information Service
        (SRIS), 8
Series network, 37–39
Shearing injuries, 49
Short circuits, 48
Software failures/errors, 78
Software safety features, 102
SRIS, *see* Safety Research Information
        Service (SRIS)
Standards, 8–9
Standby system, 43–45
Start-up safety-related factors, 176–177

Steady-state probability expressions, 193
Straining-and-spraining injuries, 49
Supervision of workers, 121
Systematic hardware faults, 78

T

Task analysis, 163
    probability tree method for, 71
Tearing injuries, 49
Technical reports, 7
Technique for human error rate
        prediction (THERP), 71
Technique of operations review
        (TOR), 68
Teleoperator for operations,
        maintenance, and construction
        advanced technology
        (TOMCAT), 149
Test-robot program method, 173
THERP, *see* Technique for human error
        rate prediction (THERP)
Time-dependent failure rate, *see* Hazard
        rate—function

TOMCAT, *see* Teleoperator for
        operations, maintenance,
        and construction advanced
        technology (TOMCAT)
Top event, 63, 85
TOR, *see* Technique of operations review
        (TOR)
Training cost (TRC), 180, 182
Training of workers, 121–122
TRC, *see* Training cost (TRC)

U

Unauthorized access, 99

W

Warning signs, 105, 110
Watchdog safety computer, 161
Weibull distribution, 22
Worker mean time to error (WMTTE),
        197, 198
Worker reliability (WR), 197
Worker–machine interface, 155